What the Printer Should Know about Ink

by
Nelson R. Eldred
and
Terry Scarlett

Graphic Arts Technical Foundation
4615 Forbes Avenue
Pittsburgh, Pennsylvania 15213-3796
United States of America
Telephone: 412/621-6941 Fax: 412/621-3049
Telex: 9103509221

Library of Congress Catalog Card Number: 90-61435
International Standard Book Number: 0-88362-143-6

Printed in the United States of America

Order No. 1311
Second Edition
Reprinted in 1995

A catalog of GATF text and reference books, Learning Modules, audiovisuals, and videotapes may obtained on request from the Graphic Arts Technical Foundation at the address given at the bottom of the title page.

Printed on
Glatfelter 60-lb. Offset Machine Finish Paper

Printing and binding compliments of
Edwards Brothers, Inc.

Contents

1. The Printing Processes 1

2. Buying Printing Ink 11

3. Color 17

4. Pigments 27

5. Flow 41

6. Vehicles 57

7. Driers 81

8. How Printing Inks Set and Dry 87

9. Ink Manufacture 97

10. Testing 107

11. Specification of Printing Inks 139

12. Lithographic Inks 145

13. Gravure Inks 181

14. Flexographic Inks 197

15. Letterpress Inks 211

16. Screen Inks 217

17. Toners and Specialty Inks 225

Glossary 231

Index 241

Printer's Dedication

Edwards Brothers' support of GATF and its programs goes all the way back to our membership in the Lithographic Technical Foundation in the 1930s, and continues up to the present day through our involvement in user groups, technical audits, and training programs. GATF has always been a valuable source of new ideas and high professional standards. We are proud to support GATF's training efforts by printing *What the Printer Should Know about Ink*, the latest in a long series of books we have printed for the Foundation.

In its 102nd year, Edwards Brothers is pleased to offer this book as thanks for the valuable work GATF has done in preparing the printing industry for the 21st century.

Martin H. Edwards
President
Edwards Brothers, Inc.

Foreword

This new edition of *What the Printer Should Know about Ink* features the latest information on ink and ink manufacturing technology. The information is presented in a basic easy-to-read format so that students as well as industry representatives can benefit from it. All of the chapters have been updated and expanded to reflect the developments that have occurred in the years since publication of the previous edition. Like the last edition, the new text covers everything from the components of printing ink to testing and specification and includes troubleshooting charts for lithographic, gravure, flexographic, letterpress, and screen inks.

What the Printer Should Know about Ink was produced through the efforts of many people both inside and outside of GATF. The text was edited by Deborah L. Stevenson, assistant publications editor, and typeset by Henry E. Grzegorczyk, phototypesetting specialist. All line illustrations were created by Mary Alice O'Toole, graphic designer.

Although all of the chapters were reviewed by the National Association of Printing Ink Manufacturers (NAPIM) and their board of technical reviewers, GATF takes full responsibility for the information presented in this text.

We give special thanks to NAPIM and its staff for coordinating the technical review of the entire manuscript. Their valuable advice, genuine interest, and overall commitment to the success of the project were greatly appreciated. We would also like to thank Peter J. McCabe of Exxon, U.S.A. for his technical expertise and willingness to review sections of the manuscript. Finally, we would like to thank Gregory Tyszka of the Gravure Association of America (GAA) and Leon Knorps of Flint Ink Corporation for updating our gravure chapter.

Thomas M. Destree
Editor in Chief

1 The Printing Processes

A printing ink is a dispersion of a colored solid (a pigment) in a liquid, and it is formulated to produce an image on a substrate. Solutions of dyes in water or other liquids are usually considered to be writing inks, not printing inks, although today ink jet printers use such materials. Some flexographic inks for special applications are also colored with dyes. Dry or liquid toners such as those used in electrostatic printing are not usually considered to be printing inks, but they are discussed in this book. In order to make the ink suitable for producing an image on a commercial printing press, other additives must be incorporated in the formula.

The printers first concern with printing inks is to get the right ink for the job. Inks that are properly formulated for one job often cause problems with another because they are not appropriate for the job at hand or because they have been altered by someone in the printing plant. Chapter 2 thoroughly discusses buying printing inks.

Because printing inks must be formulated to carry out specific jobs, attention must be paid to the printing process for which the ink is designed. This chapter discusses the printing processes, the typical products that they produce, and how both affect ink requirements.

Six basic processes for printing are listed in the accompanying table together with twelve examples of the different methods. The six basic processes are lithography, gravure, flexography, letterpress, screen printing, and nonimpact printing. There are also other specialty processes that are used in some limited applications.

Printing processes	Process	Method	Examples
	Planographic	Printing from flat surface	Lithography Collotype
	Intaglio	Printing from recessed surface	Rotogravure Steel die
	Relief	Printing from raised surface	Letterpress Flexography
	Stencil	Ink forced through a stencil	Screen printing Mimeograph
	Electrostatic	Image carried electrostatically	Xerography Electrofax Laser printing
	Ink jet	Ink droplets	Ink jet printing

There are variations within these six major printing processes. Printing may be done either directly onto the paper (or other substrate) or by offsetting from an intermediate surface, called a blanket. Lithographic printing is most commonly done by the offset process, although direct lithography (printing directly onto the paper from the lithographic plate) is sometimes used in printing newspapers. This is called Di-Litho. While usually printed directly onto the substrate, letterpress is sometimes printed offset onto paperboard in folding carton printing. The term "offset" can be used in different ways, yet, in practice, offset printing is almost always thought of as being offset lithography. Such terms as "offset," "lithography," "dry offset," and "letterset" are actually more descriptive even though they might sound ambiguous.

All offset processes have the advantage of printing smoothly and uniformly on uneven surfaces. Textured papers are best printed offset. Offset printing also produces more uniform printing on heavy paperboard, which is apt to be nonuniform.

One of the major features affecting the printing processes and the ink formulations for them is the characteristic of the ink film thickness being applied. A thin film of ink requires a higher level of pigmentation (higher color strength) than a thick film. However, thick ink films may actually contain fillers such as clay. In fact, it is often necessary to incorporate a filler into screen inks and letterpress inks—in order to achieve proper flow.

Although the printed ink film thickness applied (by any process) can vary greatly, the following numbers can be considered typical.

	Process	mils	mm	microns
Typical ink film thickness (wet film, full solid, coated paper)	Sheetfed offset	0.2	0.005	5.0
	Web offset	0.3	0.008	7.5
	Web letterpress	0.4	0.010	10
	Gravure	1.2	0.030	30
	Screen	1.0–5.0	0.025–0.125	25–125

Lithography   Lithography is now the leading method for publication and commercial printing as well as an important method for printing packaging materials. A newer process than letterpress, it was invented by Alois Senefelder in the eighteenth century and has undergone constant change throughout the years.

Lithographic printing was first done from blocks of fine, porous limestone. Lithography now prints from plates commonly made of aluminum, although plates for offset duplicators are often made from specially treated paper or plastic. The nonimage area of the plate is hydrophilic (that is, it has an affinity for water) and, when wet, will not accept ink. The image area, on the other hand, is hydrophobic (it does not accept water) and is wetted by the ink, which sticks to the image.

To create the most common type of lithographic plate, a sheet of aluminum is first coated with a light-sensitive material, usually a diazo or photopolymer. After the plate is exposed to light through a film (negative or positive), the soluble coating on the nonimage area is washed away. For positive-working plates, the exposed coating becomes soluble and is washed away; for negative-working plates, the exposed coating becomes insoluble and the unexposed coating is washed away.

During printing, the plate is first moistened by dampening solution, which is carried on a roller called the dampening roller. The dampening solution, primarily a mixture of water, gum, and acid, wets the plate. The inking rollers then contact the plate, transferring ink to the image area. Drops of water that may be on the image are squeezed off by the ink roller or emulsified into the ink, but the water that is adsorbed by the nonimage area prevents transfer of ink to that part of the plate.

Offset lithographic printing

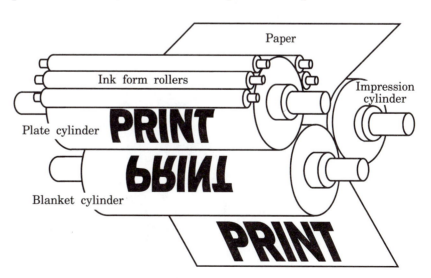

Offset was first used to improve the quality of print on uncoated paper, which has a rough surface. The rubber blanket helped in printing an even image on an uneven substrate.

Sheetfed offset lithography, the method in which single sheets of paper are printed one at a time, is used for general commercial printing, short-run publications, folding cartons, labels, and metal decorating (printing on flat sheets of steel). Web offset lithography is used for longer-run books, magazines, catalogs, newspapers, newspaper inserts, and business forms.

In web offset printing, a continuous roll of paper goes through the press. For publication printing, both sides of the paper are printed simultaneously between two blanket cylinders, each cylinder serving as the impression cylinder for the other. When both sides of the paper are printed during a single pass through the press, the operation is called "perfecting."

Gravure

In gravure printing, the cylinder has a copper surface that is either chemically etched or mechanically engraved to form the image. The size and depth of each halftone dot can be modified to give gravure the widest tonal range of all printing processes. When the press is running, the cylinder rotates in the ink fountain, and a doctor blade made of metal wipes the excess ink from the surface of the plate. The cylinder is then pressed against the substrate, and the ink in the cells transfers directly to the material.

Rotogravure printing

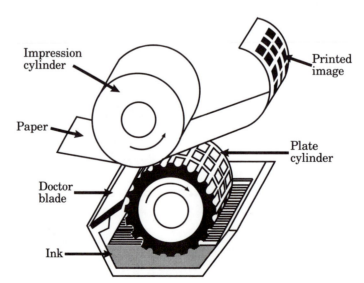

Because the inking system is so simple and because the inks can be made with high-speed solvents, the gravure process is able to print at very high speeds permitting the ink to dry at low temperatures. However, the inks can be highly volatile.

Products commonly printed by gravure are long-run magazines, catalogs, labels, newspaper inserts, Sunday supplements, flexible packaging, carton board, and specialty materials like wallpaper and decorative panels.

The term "intaglio" is sometimes used to refer to all printing from engraved plates (including rotogravure), but today it generally refers only to special applications such as steel die engraving. United States currency, treasury bonds and notes, and some postage stamps are printed with a copper plate or a steel die engraving. Copper plates for this type of printing may be hand-engraved, but steel plates are engraved from a hardened die. The engraved plate is first filled with ink, then excess ink is cleaned from the surface, and the plate is pressed onto the paper to be printed. The process is capable of extremely high-quality line work.

Flexography

Flexography is the principal method of relief printing, or printing from a raised image. The roller configuration, the type of ink, and the type of plate make flexography considerably different from conventional letterpress. Flexography (flexo) uses flexible rubber or photopolymer plates. The rubber plates are molded in hydraulic presses against a metal or hard plastic matrix. The photopolymer plates, however, are made by exposing the plates to light through a photographic film and then washing away the unexposed coating.

Four-cylinder flexo unit

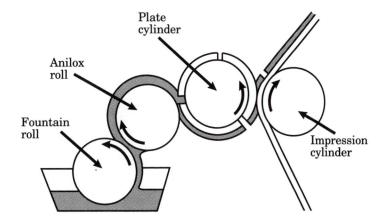

Plate cylinder

Anilox roll

Fountain roll

Impression cylinder

Three-cylinder
flexo unit

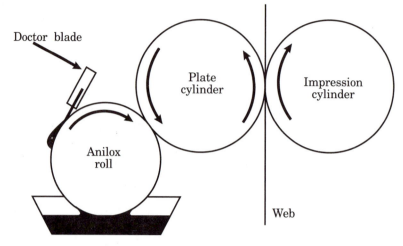

Flexography uses a low-viscosity (liquid) ink (instead of a paste ink as letterpress does), and consequently the inking system of the flexo press is very different. Web letterpress equipment has as many as nineteen rollers in the inking train and therefore must use relatively stable inks that dry either in a heatset drying oven or by absorption. In flexography, only two or three rollers are used. Flexographic inks are much more volatile and are usually dried with high-velocity warm air. This makes the flexo process particularly suitable for printing on plastics that melt at low temperatures. These plastics are used for many types of flexible packaging.

Once confined to relatively low-quality printing, technical advances in flexography have made it capable of producing excellent four-color process printing and halftone reproduction. It is used on such substrates as cellophane, polyethylene, polypropylene, glassine, carton board, and corrugated board.

Flexography is widely used for printing flexible packaging, corrugated boxes, folding cartons, paper bags, paper and plastic containers, and narrow-web labels. It is also being used increasingly to print newspapers.

Letterpress

Letterpress is one of the oldest printing processes. Long before Gutenberg developed a system of movable type, woodcut prints were used to produce graphic designs in Asia as early as the ninth century. Letterpress prints from a raised image area, which means it is a relief printing process. The raised image area is then pressed against a substrate such as paper or paperboard.

Letterpress
impression

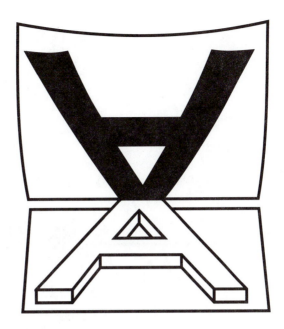

Offset letterpress printing, sometimes called "letterset" or "dry offset" is used when the substrate to be printed is hard and impervious such as wood, plastic, or metal. Two-piece beverage cans are printed by dry offset on high-speed cylindrical presses. Some can makers are using the dry-offset process with ultraviolet inks.

Although letterpress printing is rapidly being displaced by lithography, flexography, and gravure, it is still done in many commercial job shops that print handbills, business cards, and other small specialty items. Letterpress is still used, but to a diminishing extent, for printing newspapers.

Screen Printing

Screen printing is a stencil printing process where ink is forced through a screen, or mesh, onto a substrate. The process was once called "silk screen" because the screens were originally made of silk. The term "screen" printing is now preferred because the mesh is usually made of metal, nylon, or polyester.

Screen printing is ideal for printing irregularly shaped objects. Posters, wallpaper, clothing and other textiles, decalcomanias, ring binder covers, instrument panels and dials, preformed metal and plastic containers, and circuit boards are among the many products printed by the screen process. Screen inks can be made to dry by evaporation, oxidative polymerization, heat curing (thermal polymerization), or ultraviolet curing.

Screen printing
process

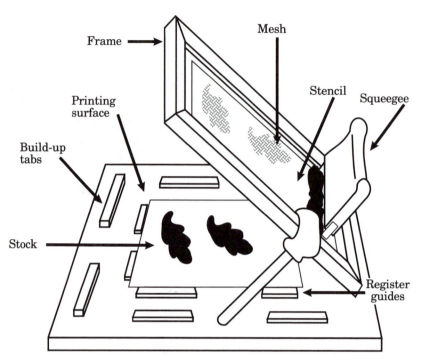

**Nonimpact
Processes**

There are many nonimpact printing methods: electrophotography (xerography and laser printing), ink jet,
magnetography, electrography or ionography, thermal
transfer, and others. Only the first two will be discussed
here.

Electrophotography. Without question, the fastest-
growing method of printing, at least on a percentage basis,
is "electrophotographic" printing, sometimes called
"xerography." In this printing method, a drum or flat
plate, usually made of selenium or a photoconductive
polymer, is charged electrostatically (in the dark). When
light, either from a reflected copy or from a computer-
controlled laser beam, hits the drum, the charge is
dissipated leaving an electrostatic image. Toner is then
applied to the drum or surface. It is retained by the
electrostatic image, creating a visual image that is then
transferred to the paper. When the paper is heated, the
toner is fused, producing a permanent copy.
  Although the imaging mechanism within the machine
itself may be complex, electrostatic printing machines
(copiers and duplicators) are the ultimate in processing
simplicity; the operator loads the machine (after putting in
toner and paper) and presses a button. Electrostatic

The electrostatic process

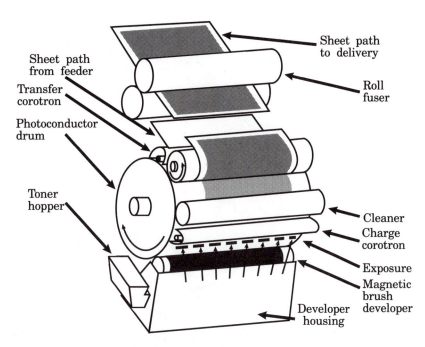

Sheet path to delivery

Sheet path from feeder

Transfer corotron

Photoconductor drum

Toner hopper

Roll fuser

Cleaner

Charge corotron

Exposure

Magnetic brush developer

Developer housing

printing, which is primarily limited to office copiers and in-plant and quick-copy printing, has become a major method of printing today. Machines printing as many as 7,500 sheets an hour are commonplace.

Toners used for electrostatic printing may be "dry toners" or "liquid toners." Dry toners are made of carbon black or another pigment blended with a thermoplastic resin that melts when heated and cools to produce a solid, permanent copy. Liquid toner is a suspension of pigment in a volatile liquid.

Ink jet printing. The ink for ink jet printing is usually a solution of a dye in water or a solvent (such as alcohol). A jet of electrostatically charged ink droplets is generated; the ink droplets, which are controlled by a computer-generated charge are projected onto the paper in the form of an image. The ink droplets may come from a continuous jet, or they may be generated on demand. The nonprinting ink droplets from the continuous jet are captured and recycled.

At high speed, the process is not capable of the high resolution that is required for commercial printing, but at low speeds, ink jet printing is used to prepare proofs of process-color printing. The process is especially suitable when only one or a limited number of copies of a large variety of images are desired. High-speed ink jet printing is used for coding, marking, addressing, and for direct-mail letters.

Digital continuous
ink jet
*Courtesy Videojet
Systems International, Inc.*

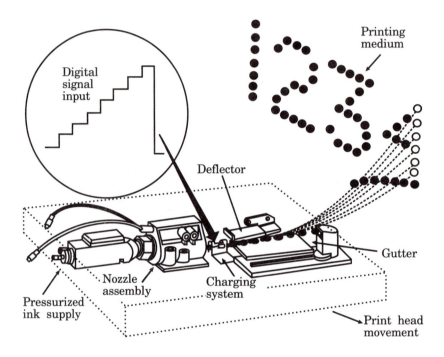

Digital signal input

Printing medium

Deflector

Gutter

Nozzle assembly

Charging system

Pressurized ink supply

Print head movement

2 Buying Printing Ink

Price vs. True Cost and Value

The true cost of printing ink is more than just its purchase price. The true cost includes the mileage of the ink, its behavior on press, end-use performance, and the ink manufacturer's service. Even if the purchase price of the ink accounts for only 2% of the price of the printed job, the printer will lose money if the job requires extra time on press, cannot be delivered on time, or is rejected because of the ink. The price is apparent on the invoice, but true cost and value are difficult to determine.

Ink manufacturing is still largely a prescription manufacturing business. Although off-the-shelf inks are often suitable for many jobs, printers should consult with their ink suppliers before purchasing any ink. Once given both the printer's and customer's requirements, the ink manufacturer will determine whether an off-the-shelf ink is appropriate for a particular job or whether a special ink must be formulated.

A well-made expensive ink may actually end up being less expensive than an ink that costs less if the printer uses less of it and gets better performance. The ink manufacturer can reduce the price of an ink by decreasing its color strength and dispersion or by manipulating it in other ways. The most expensive ingredient in the formulation is usually the pigment. If the pigment concentration is reduced, the ink looks the same in the can, but the printer has to run more of it to bring the print up to color.

Ink Mileage

It is always difficult to estimate the number of thousand square inches that a pound of ink will cover (or, in metric units, the number of square meters covered by a kilogram). Ink consumption is determined by the color strength of the ink, its mileage, the type of paper used, and the care that the press crew takes.

The color strength of the ink is one of its most important properties, and one that can always be reduced to lower the price of the ink. The printer can run a bleach-out test (as described in Chapter 10) or send the ink to GATF or another ink laboratory to have the color strength determined. Other factors being equal, the ratio between the color strength of two inks will be the same as the mileage to be expected—the greater the color strength, the greater the ink mileage. The kind of paper used for the job also determines the amount of ink needed. An enamel stock requires less ink (yields better ink mileage) than a dull coated, which requires less ink than an uncoated.

Ink consumption also depends on the care that the crew takes when handling it. If makeready is well planned and fast, less ink is used.

Finally, the amount of ink to be ordered depends on the skill of the estimator. Even experienced estimators differ in estimating the amount of ink for a complete job. Printers can consult estimating charts for a rough idea of how much ink they need, but some of them differ greatly and most of them apply only to sheetfed offset.

Securing the Best Supplier

Printers should attempt to form close relationships with one or two suppliers who are willing to meet their needs and provide them with quality ink. It is always wise to have a backup supplier in case the primary manufacturer is not able to fill an order. Attempting to secure more than two or three suppliers is discouraged because it is not possible to give them enough business to make a close, attentive relationship worthwhile and because buying ink from several different manufacturers generates waste and unnecessary storage and inventory costs.

Cooperation with suppliers comes through the mutual effort of both parties. Ink companies will only service the inks that they sell, and the printer must be satisfied with the required service that the supplier is willing and able to provide. The printer should not hesitate to ask for the service included in the price of the ink, but, by the same token, if the printer asks for unreasonable service, the ink manufacturer must find some other way to maintain profits. Ink suppliers who know and understand the printer's customer can provide the printer with optimal service.

Basic Goals in Buying Ink

GATF has observed that the biggest problems with ink result from an ink not being suitable for a particular job. This can result when the printer does not give the sales representative complete and correct information, or the sales representative does not ask for it; the printer picks a can off the shelf without concern for runnability and end use; the ink supplier makes a mistake; or someone at the printing plant alters the ink after it has been properly formulated.

It is not necessary to ask for a new ink formulation every time the same job is run, but the first time a different substrate is used—carton board, poly-coated board, matte-coated, cast-coated, or some other new or different substrate—the printer should consult the ink manufacturer.

Items to be considered when ordering ink

Information	Comment or Example
Type of press	Offset, flexo, or gravure; sheetfed, sheetfed IR, sheetfed UV; web, web offset heatset, web offset nonheatset, web offset EB, web offset UV; single-color or multicolor
Ink drying system	High-velocity hot air; absorption by newsprint
Drying time and drying requirements	
Running speed	
Required color	Submit specimen for color match
Color sequence	Order of printing colors
Tack and tack sequence	
Type of substrate	Paper or board, uncoated, enamel, matte, cast-coated; plastic or plastic-coated paper or board; metal; foil; film; glass
Desired finish	Gloss, matte
Dampening solution to be used	
End-use requirements	Product resistance, lightfastness, aging, weather resistance, rub resistance
Processing and converting requirements	
Safety or environmental considerations	
Ink mileage	Run length and amount required
Delivery requirements	Suitable package size; delivery deadline
Price	

If ink comes out of the can press-ready, costs of modifying the ink and wastage from "fiddling around" are eliminated. As unbelievable as it may seem, things like motor oil, axle grease, vegetable oil, corn starch, lard, cream of tartar, etc. are still being added to the ink. These materials should **never** be added to printing ink. If the ink doesn't run right, consult the ink manufacturer.

The printer and ink sales representative should discuss potential problems before they arise so that the proper additives will be on hand if they are needed. For instance, on small jobs it is sometimes helpful to add drier or reducer to an ink to help overcome a press problem. The printer should know what type of drier to add to the ink before the problem arises.

Purchase ink the most economical way. Ink prices vary with the size of the kit or container. Returning ink to the can from the ink fountain takes time and increases the possibility of contamination, but throwing unused ink away may be even more costly. The ink supplier can help the printer get the most out of the ink by specifying the most economical size of container and means of shipping.

The printer should buy ink in the largest containers that the plant can use efficiently. Ink in kits costs less than ink in cans, and ink in drums costs less than ink in kits. Although tote bins represent a large investment and returning them to the ink manufacturer adds shipping charges, they are often most economical for the printer who uses large quantities of ink.

If the printer is big enough to utilize the output of a small ink plant, the ink supplier can sometimes set up an ink blending facility within the printer's plant. This reduces shipping, packaging, and inventory costs and allows for close communications between the printer and ink manufacturer. An ink technologist within the printing plant can make any required ink alterations.

Disposal of empty printing ink containers. Many states (including Pennsylvania and New Jersey) have very stringent industrial waste (nonhazardous) disposal regulations, which may affect the disposal of empty printing ink containers (pails and cans).

Ink storage area. The ink storage area varies depending on the type and amount of ink being handled. These two variables are determined by the type of press and printed product. For instance, ink storage and handling differ vastly between a sheetfed pressroom, a web offset commercial pressroom, a gravure pressroom, and a newspaper pressroom. When building a new plant or renovating an old one, the printer must get help and advice from the ink supplier and other competent consultants. Proper design of the ink room can prevent endless problems later.

The design of the ink storage room must allow for convenient delivery of ink from receiving and to the press. Inks must be stored at temperatures close to that of the press because of the great effects of temperature on ink flow to the press and on the press. Water-based inks, especially, must always be protected from freezing.

Safe handling of printing ink. Printing ink is a mixture of chemical compounds, and many safety and health regulations are concerned with the safe handling of chemicals and the disposal of chemical wastes.

Under certain conditions, inks and other chemicals used in printing can cause safety and health problems for the people who are exposed to them. Manufacturers are required by law to provide Material Safety Data Sheets (MSDSs) that describe safe use, storage, and emergency handling of their products.

Some MSDSs contain warnings not present on a product label. Before a new kind of printing ink is used, the technical manager should review the MSDS and product label to evaluate the potential hazards posed by the ink components and, if necessary, provide training for employees as required under the OSHA Hazard Communication Standard.

Establish inventory goals. Printers cannot afford to postpone a job until the ink arrives or to run out of ink during a run. To avoid this, printers should establish an inventory system in which they set maximum and minimum storage goals based on average monthly consumption. An early warning system and enough flexibility to accommodate a big rush job that will consume an unusually large volume of ink are required. Printers should also make sure that reorder levels are high enough to give the ink supplier time to manufacture and deliver a new batch.

In implementing an inventory system, a first-in, first-out policy is a must. Placing new ink at the rear of the storage area ensures that old ink will be used first. Most inks are expected to retain their intended properties for approximately a year. However, changes in the formulation do occur over longer periods of time and old inks should be discarded to avoid spoiling a print job.

Printers should also standardize their ink rooms, storing as few special colors as possible. Even though different sets of process colors are required for different substrates, the number of sets should be kept to a minimum to eliminate clutter and possible waste.

3 Color

Color science is a vast subject including both physics and psychology. A comprehensive discussion of color is beyond the scope of this book. This chapter will give printers a few fundamentals of color that will help them improve the consistency and predictability of color reproduction. Those who wish greater detail should consult GATF's *Color and Its Reproduction* written by Gary G. Field.

Appearance and Measurement of Color

Both instruments and the eye are used to control color in the printing plant. Instruments are necessary because color control requires objective, numerical measurements. With the human eye, color perception varies from person to person and people have poor color memory. On the other hand, the eye not only can observe the entire sheet or signature, it is also an excellent comparator—that is, it can detect very small differences in color.

Illumination

Light is a type of electromagnetic radiation that can be seen by the eye. Other electromagnetic waves include X rays, ultraviolet and infrared radiation, and radio waves.

The wavelength of light is measured in nanometers (nm) or millimicrons. (One nanometer is one-thousandth of a millionth of a meter, or 1×10^{-9} meter.) An infrared ray with a wavelength of 1,000 nm has a wavelength of one micrometer (micron) or one-millionth of a meter (about 0.00004 in.).

It would seem obvious that the apparent color of a print is affected by the color of the light used for viewing, but it is astonishing how often this simple fact is overlooked. A print that appears one way in the pressroom can look different on the superintendent's desk and appear yet another way in the customer's studio.

Printer, supplier, and customer must agree on the illumination under which the print is to be viewed if they are to agree on the results. International standards of illumination are usually most satisfactory because they are readily available, carefully specified, and generally agreed upon. The printer should be thoroughly familiar with "Viewing Conditions for the Appraisal of Color Quality and Color Uniformity in the Graphic Arts," International Standards Organization (ISO) standard no. 1349/A and American National Standards Institute (ANSI) standard no. PH2.32-1972. (Test methods are discussed in Chapter 10.) Among other things, this standard requires 7,500 K illumination for comparison between press sheets and 5,000 K for comparing press sheets with originals. Most printers, however, prefer to use a single level of 5,000 K for both comparisons.

Metamerism is the name given to the phenomenon of printed colors matching under one light but not under a different light. It is apt to occur whenever an ink has been blended or mixed to match a desired color. Since there are dozens of ways of matching any given color, it is highly unlikely that the colors will match under all kinds of illumination.

Measurement of Color

Methods of measuring color are usually classified into three groups: densitometry, colorimetry, and spectrophotometry. In the printing plant, densitometry is satisfactory for most purposes. The densitometer measures optical density. Optical density is a measurement of a material's ability to absorb light. It is proportional to the amount of light reflected from or transmitted through a sample. (A reflectometer records or reads the percent reflectance of a beam of light compared to the standard beam.) Transmission densitometers compare transmitted light, and reflection densitometers compare reflected light.

Transmission densitometer with printer
Courtesy Cosar Corporation

Because different makes of densitometers may vary in regard to spectral response, the printer should choose one brand of densitometer as the standard and use it throughout the company.

Microprocessor-enhanced densitometer
Courtesy Brumac Industries

Most densitometers used in the printing plant are equipped with three color filters—red, green, and blue. The blue filter measures the reflected light in a range between about 400 nm and 500 nm and reads it as the blue density. Similarly, the green filter measures the reflected light in a range between about 500 nm and 600 nm and the red filter measures in a range between about 600 nm and 700 nm. The exact range of wavelengths measured depends on the specifications of the filters used with the densitometers. For example, so-called wide-band filters typically measure

Reflection densitometer with printer
Courtesy Cosar Corporation

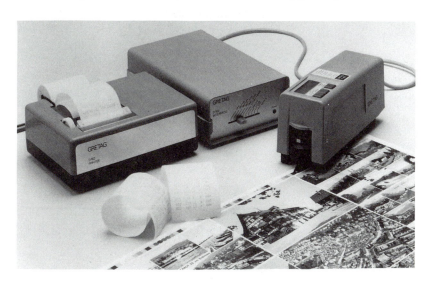

a range of 50–75 nm, while the "narrow-band" filters generally measure a range of about 20 nm. With both types of filters, the density reading is an average of the densities over the range of wavelengths measured, not the density at a specific wavelength. The following table shows typical densitometer readings for an average magenta ink film printed on a paper substrate.

Densitometer readings of typical magenta (rhodamine)

	Optical Density	Percent Reflectance*
Red filter	0.12	76
Green filter	1.30	5
Blue filter	0.68	21

*Calculated from optical density

The spectrophotometer measures the reflectance of each individual wavelength. It gives the entire spectral reflectance curve instead of only average reflectances over three broad wavelength regions, as a densitometer would.

A colorimeter measures color in a very different way. It gives three numbers, which represent hue, intensity, and value. The hue is the basic color (e.g., orange, green, or blue). The intensity, also called chroma, saturation, or color strength, refers to the concentration or dilution (attenuation) of the color. Value is the lightness or darkness of the color. Each number can vary independently. A red can be strong and dirty, such as brick red, or strong and clean, such as a stop sign, or weak and clean, such as pink.

A popular method of measuring and recording color is the L,a,b scheme developed by Hunter Associates Laboratory in

Hunter L,a,b chromaticity diagram

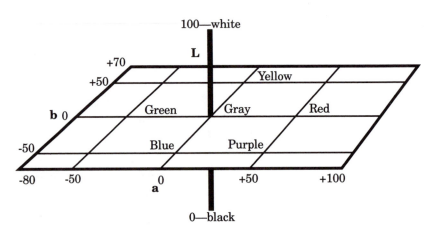

1947. The "L" number refers to the lightness of the color; for example, for white, L=100, for black, L=0. The "a" denotes the degree of redness or greenness; if it is positive, the color is more red, and if it is negative, the color is more green. The "b" value indicates the degree of yellowness and blueness; positive values denote more yellow and negative values more blue.

Additive and Subtractive Colors

White is defined as the presence of all colors, black as the absence of all colors. When red, green, and blue light are blended equally (as on the TV screen), white is the resulting color. When red, green, and blue are masked out, as on a printed sheet, the resulting color is black.

At first glance, it would seem that when a sheet is printed, something (ink) is added, but what happens is much easier to understand if one considers that ink subtracts color: it acts as a color filter. Black filters out all colors, yellow filters out blue, magenta filters out green, and cyan filters out red.

A "perfect" or "pure" magenta ink printed on a perfect substrate would absorb all of the green light and none of the red or blue.

Spectral reflectance curves of a "perfect" magenta ink *(left)* and a typical magenta ink

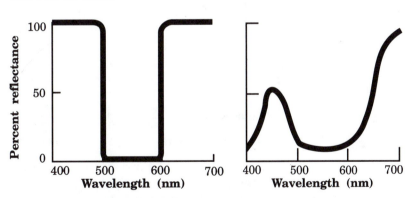

To produce a blue color, inks that filter out the red and green must be used—the cyan and magenta inks. In practice, none of the process inks are perfect: the magenta usually filters out too much blue (which is why magenta inks are too red), and cyan inks filter out some of the green and blue and are, therefore, too gray.

Yellow pigments usually come closer to doing the required job of filtering out one color and transmitting the other two than do cyan or magenta inks. This is why GATF urges the use of the terms cyan and magenta instead of "process blue" or "process red" and strictly

Spectral reflectance curves of a "perfect" cyan ink *(left)* and a typical cyan ink

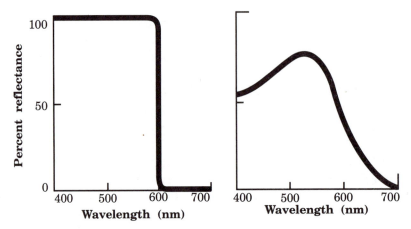

avoids the use of "blue" for cyan or "red" for magenta. Blue and cyan are very different colors, as are red and magenta, and using the names interchangeably or carelessly has resulted in embarrassing and very costly mistakes for some printers.

Spectral reflectance curves of a "perfect" yellow ink *(left)* and a yellow ink

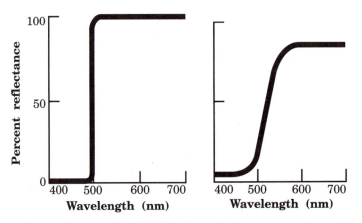

The printer should order a set of process colors from the ink supplier—four inks that are properly matched to produce good overprint colors. When using process color, the printer should use the least possible number of sets—not more than one set for each substrate used: one for enamel, one for uncoated, and one for dull-coated, as recommended by the ink manufacturer. Maintaining more than the minimum number increases the cost of inventory and creates additional variables in the pressroom.

Color Matching

Altering the ink on hand to yield the desired color is an extremely complex operation and requires a great deal of skill. The printer is urged to order the right ink color in the first place. The ink manufacturer employs skilled color matchers, craftspeople who not only understand the physics of color and the chemistry of ink, but who also have the right equipment and ink formulas so that they know which additives will be compatible and which ones will not be.

Unfortunately, the above sound advice is sometimes impractical; for example, a special job may require a few ounces or a few hundred grams of a designated spot color. Ink manufacturers normally charge extra for mixing small quantities, and there may be a delivery delay, so that the most convenient way may not always be practical.

Most ink manufacturers sell a series of colored blending inks with directions for mixing them and a swatch book showing the mixed colors printed on a card stock and identified by number. If the printer carefully follows directions and uses clean equipment, satisfactory spot colors can be produced.

PANTONE MATCHING SYSTEM® *. The PANTONE MATCHING SYSTEM was introduced by Pantone, Inc. in 1963. The PANTONE Basic Colors—PANTONE Yellow, PANTONE Warm Red, PANTONE Rubine Red, PANTONE Rhodamine Red, PANTONE Purple, PANTONE Reflex Blue, PANTONE Process Blue, PANTONE Green, and PANTONE Violet—are used together with PANTONE Black and PANTONE Transparent White to produce 724 different colors, each identified by a PANTONE Number and reproduced in a color-matching book that can be used to select the desired color. PANTONE Colors also feature fluorescent and metallic inks. Pantone supplies the formula to produce the color selected by the customer or artist. The inks are manufactured by ink manufacturers under license to Pantone, which monitors the PANTONE Basic Color inks to make sure that the ink manufacturer supplies the correct color.

American Newspaper Publishers Association. Ink colors for newspapers are specified according to AdPro or AD-LITHO, the trademarked names for ANPA's letterpress and lithographic color systems for newspapers. The next

---

*Pantone, Inc.'s check-standard trademark for color reproduction and color reproduction materials.

ANPA-COLOR book will contain recommendations for flexo colors. The color swatch book gives the blending formulas for some sixty mixes, based on eight or ten standard colors, that are reproduced in the color book. The advertising agency specifies the color, and the newspaper can get the properly formulated ink from its ink manufacturer or do its own mixing.

These systems have greatly reduced color variations in ads run nationwide. Most newspapers that run color use this color control system.

Instrumental systems. Pantone has also developed the PANTONE Color Data System, which uses computers to select and print out specifications for individual customized color matches. A small, dedicated spectrophotometer in the printing plant does the calculations. The ACS system is supplied by Applied Color System, Inc. As with the PANTONE Color Data System, the ACS employs a spectrophotometer to scan the color to be matched. The data are fed into a computer that provides a formula to match a target color or to correct an off-shade match. The computer can also calculate additions required to work off waste ink by mixing it with fresh ink to produce the desired match.

Mixing inks in the printing plant. Ink manufacturers are usually willing to set up in-plant operations in large print shops for which in-plant ink manufacture is worthwhile. For smaller shops, consultants can show any printer how to blend and mix inks. However, the smaller printer must rely on this consultant for the assistance and advice formerly obtained from the manufacturer and should carefully weigh the merits of the two ways of obtaining printing ink.

Color Variation

One of the biggest problems of color printing is the failure of the printing system to produce the desired color consistently. Since color may vary during the run, the color of the print may fail to match the original. Besides dot gain, trap, and inconsistent ink color strength, paper differences are among the most common causes of color variation. The color of the paper as well as its finish affect the color of the printed ink film. A "warm" sheet, which has a relatively high reflectance of light in the red or orange region of the spectrum, will not print the same color as a "bright" sheet, which will have a high reflectance of blue light. This is why printers often choose a "white" sheet for printing—a sheet that diffusely reflects light of all wavelengths throughout the visible spectrum. A blue sheet, which absorbs red light, cannot print a good red color.

A paper's gloss and absorbency also affect print color. Two papers with identical color may produce very different prints if they have different gloss or absorbencies. Paper properties and their relationship to color are discussed in greater detail in GATF Research Progress Report 60, *A New Method of Rating The Efficiency of Paper for Color Reproductions.*

Dot gain is another leading cause of color variation. If dot gain (or dot loss) occurs in prepress so that the dot value on the printing plate or cylinder is incorrect for the amount of color required, the color may be wrong, but at least it will be consistently wrong. When dot size changes on the press as a result of variations in ink feed or printing pressure, these changes often vary during the run and produce inconsistent color.

Ink trap, a third cause of color variation, is discussed in Chapter 5. Variations in the thickness of a second-down ink film have important influence on color. The color tends to vary from print to print. Excellent consistency requires careful formulation of the ink, and altering the ink in the pressroom is one cause of poor trapping. Changing ink feed to alter the optical density of the printed ink film obviously changes the color of the print, but variations of optical density cause less trouble than variations in dot gain and trapping. Accordingly, it is difficult for the press crew to overcome dot gain and trapping by changing the ink feed.

Good color consistency requires care throughout the operation—careful and adept prepress work and presswork and the appropriate paper and ink. Because these change so often, it is apparent why color variation is a major printing problem.

Undercolor Removal and Gray Component Replacement

Certain combinations of yellow, magenta, and cyan yield blacks and dark grays. Undercolor removal (UCR) is incorporated into many color separations in order to reduce the yellow, magenta, and cyan dot values wherever black is going to print. In other words, color is removed from the neutral scale. Instead of printing 100% values of yellow, magenta, cyan, and black in a certain area, a set of UCR separations might print, say, 60% yellow, 60% magenta, 70% cyan, and 70% black in that same area.

Gray component replacement

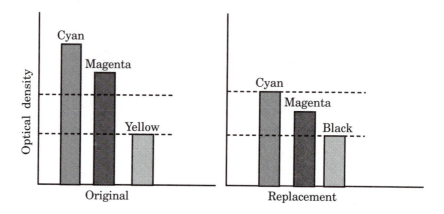

Whenever dots of yellow, magenta, and cyan are present in the same color, there is a gray component to that color. In gray component replacement (GCR), the smallest of the three dot values is removed from the color, together with appropriate amounts of the other colors in order to produce a neutral tone, and then that gray tone is replaced by an appropriately sized dot of black.

UCR and GCR offer two important advantages. In GCR, costly process inks are replaced with less-expensive black, and in both processes, not as much ink is needed. Applying less ink to the paper reduces setoff problems in sheetfed printing and ink drying problems in web printing.

GCR cannot be initiated in the pressroom. It requires complicated computerized calculations in the scanner-generated color separations. The press crew must use the plates as they are provided, exactly as with any other printing operation.

4 Pigments

Pigments are finely ground solid materials that impart color to inks. The nature and amount of pigment that an ink contains, as well as the type of vehicle, contribute to the ink's body and working properties. A century ago, most pigments were colored minerals ground to fine powders. They were extremely fast to light, water, and solvents, but opaque, weak in color strength, and very dull. Even when finely ground, they were still gritty enough to cause rapid wear of printing plates. Lampblack, made from natural gas, was the principal black. Other pigments were made from vegetable dyes, such as indigo, madder, and logwood extract, and from carmine dye derived from the cochineal insect. Only a few synthetic pigments were available. Over the years, the texture of pigments has been softened, and their vehicle-wetting properties and uniformity improved.

The synthetic dye industry originated in the second half of the nineteenth century due to advances in organic chemistry. Its growth led to the development of organic pigments suitable for inks. Organic pigments are usually more transparent, brighter, purer, and richer in color than their inorganic counterparts. Developments in chemistry have also led to the production of better inorganic pigments such as titanium dioxide.

Pigment Properties

A pigment may have many attractive properties and yet not make a good ink; e.g., phloxine is a relatively inexpensive magenta pigment with an excellent shade, but it contains lead, bleeds in alcohol, and the color fades rapidly. Pigment characteristics of importance in ink formulation include tinctorial strength, opacity, shade, gloss, durability, particle size, specific gravity, refractive index, hardness or texture, wettability, dispersibility, and light, heat, and chemical resistance.

Particle Size

The **particle size** of ink pigments ranges from about 0.01 to 0.5 microns. (Since a micron—one one-millionth of a meter—is approximately 0.00004 in., these values correspond to 0.0000004–0.00002 in., also written as $4 \times 10^{-7} - 2 \times 10^{-5}$ in.) Carbon blacks have the smallest particle size. They usually exhibit excellent flow properties. Coarse pigments, like titanium dioxide, generally produce opaque inks, which tend to settle out, and can pile on the ink rollers, plates, and blankets. Special formulations are often necessary to make such inks transfer properly.

Pigment properties

C.I. Pigment Name	No.	Name	Color	Dispersibility	Brightness	Color Strength	Resistance to Water/Alcohol		Lightfastness (wax scale)	Lightfastness Interior Full	Interior Tint	Exterior Full	Exterior Tint	Relative Price
Yellow 1	11680	Hansa G	Yellow	E-G	G	G	E	P	6	E	F	G-F	P	20
Yellow 12	21090	Diarylide yellow AAA	Yellow	—	E-G	E-G	E	E-G	3	F	F-P	P	P	22
Yellow 13	21100	Diarylide yellow AAMX	Reddish yellow	—	E-G	E	E	E	3.5	F	F	P	P	30
Yellow 14	21095	Diarylide yellow AAOT	Greenish yellow	—	E-G	G	E	E	3	F	F-P	P	P	25
Red 53	11585	Red lake C	Red	E	—	G	P	P	2	P	P	P	P	—
Red 53:1	11585	Red lake C (Ba salt)	Red	G-F	—	G	E	F	—	F	P	P	P	15
Red 57	15850	Lithol rubine	Bluish red	E-G	E-G	E-G	E-G	G-F	3	G	G-F	G-F	F	15
Red 90	45380	Phloxine	Bluish red	G	E	—	G-F	P	1.5	P	P	P	P	20
Violet 1	45170	Rhodamine B	Violet	—	E	E	G	P	3	G-F	G-F	—	—	40
Violet 3	42535	PMTA violet	Violet	—	—	E	E-G	G	3	F	F	F	F	40
Blue 15	74160	Phthalocyanine blue	Blue-green	G-F	E	G-F	E	E	7.5	E	E	E	E	32
White 6	77891	Titanium dioxide	White	F-P	E	G	E	E	—	E	E	E	E	27
Black 7	77266	Carbon black	Black	F-P	—	E	E	E	—	E	E	E	E	—

E = Excellent, G = Good, F = Fair, P = Poor

Source: *Pigment Handbook*

Specific Gravity

The **specific gravity** of a pigment is the ratio of the weight of a given volume of the pigment to the weight of the same volume of water. The specific gravity of inorganic pigments is higher than that of organic pigments.

Refractive Index

The **refractive index** is a measure of the bending (or refraction) of light rays entering the pigment. If the pigment has the same refractive index as the vehicle in which it is suspended, the light rays will pass through the ink without being bent, and the ink film will be more or less transparent. If the refractive indexes are different, the light rays will be bent and scattered as they pass through the pigment, and the ink film will be more or less opaque; the greater this difference, the more opaque the ink film. The refractive indexes of pigments vary much more than the refractive indexes of vehicles. Inorganic pigments, such as titanium dioxide and the chrome yellows, have high refractive indexes. Chrome yellows, however, contain lead and will be used less and less because of environmental restrictions.

Opacity

Opacity is the degree to which an object (ink) bars light. How an object reflects or scatters light determines its opacity. A particle size of about 0.2–0.4 micron, or about half the wavelength of light, is necessary for maximum opacity. Carbon black pigments, which are very fine, are opaque because they absorb light.

Wettability

Wettability refers to the ease with which pigments can be wet by the ink vehicle. Dry pigment particles are always covered with a film of absorbed moisture that must be displaced by the varnish in inkmaking. Complete pigment wetting is often necessary with offset inks to prevent breakdown and emulsification in dampening solutions. In recent years, much progress has been made in the surface treatment of pigments to increase their wettability by ink vehicles.

Dispersibility

In addition to being wettable, pigment particles must also be **dispersible.** Dispersion of the pigment is necessary because it enables the vehicle to separate and surround the particles. Good dispersion requires high shear and also depends on the wettability and strength of the pigment granules, among other things.

Texture

Texture is the hardness or softness of a pigment in its dry form. If it rubs out easily to a smooth, soft powder between finger and thumb, it is soft. If the powder feels gritty, it is hard.

Lightfastness

Lightfastness depends not only on the chemical nature of the pigment and the thickness of the printed ink film but also on the following:
- Concentration of the pigment in the ink
- Protective properties of the ink vehicle
- Time of exposure
- Atmospheric conditions during exposure
- Light intensity during exposure

Lack of lightfastness does not always result in fading. In some cases, it results in darkening or some other color change. The lightfastness of paper is important. If the paper darkens or turns yellow on exposure, it will alter the print colors, especially the tints. Because the conditions of use that affect colorfastness are so many and varied, ink manufacturers usually refuse to guarantee results. They should, however, be able to give valuable advice, and printers should consult them when the lightfastness of an ink is particularly important.

Most pigment properties are affected, sometimes very significantly, by variations in the manufacturing process. The particular salt used to precipitate the pigment greatly affects solubility and often affects the shade. Resin treatment significantly alters the dispersibility. A complete description of pigments, therefore, goes far beyond what the printer usually needs to know. Because pigments vary so much in particle size, specific gravity, and wettability, each pigment requires individual consideration in ink formulation. The nature of individual pigments is described in the following sections.

Manufacture of Pigments

Pigments used in paint, cosmetics, printing inks, textiles, plastics, and other applications are made by pigment manufacturers.

Organic pigments are often made from petroleum: blacks are made by burning gas or oil; colors are made by reacting organic chemicals derived from petroleum. Phthalocyanines are prepared by mixing and heating the ingredients in a solvent with a high boiling point. Organic pigments used in printing ink represent about 65–70% of the volume and 45–50% of the value of all organic pigment consumed in the U.S.

Inorganic pigments are sometimes mined and purified, but they are usually produced through chemical reactions. Inorganic pigments are often synthesized merely by mixing aqueous solutions of selected chemicals, filtering off the precipitated pigment, and washing and drying it. Sometimes the reactants are mixed dry and calcined (heated). Use of inorganic pigments in printing inks is basically limited to titanium dioxide (TiO_2), lead chromate, molybdate orange, iron blue, and certain iron oxides.

Azo pigments are prepared by reacting two chemicals (both often off-white) in water to make an intensely colored solution. If the product is soluble in water, as it often is, another chemical is added to make the dye insoluble, that is, to convert it to a pigment. To make this wet slurry of pigment suspended in water into a useful product, the pigment manufacturer filters and collects the wet pigment, pressing out as much water as possible.

Pigment Flush Pigments are **flushed** when a wet filter cake (sometimes referred to as "press cake") is put into a mixer with a selected varnish and then processed. The wet pigment cake becomes preferentially wet with the varnish and transfers from the water to the varnish. When the process is finished, the water is poured off. Any additional moisture is removed by vacuum and heat.

Flushed pigments are not only easily dispersed into printing ink formulations, they are also, overall, less costly to the ink manufacturer than dried pigments.

Inks made from flushes have high color strength and are free from gritty pigment particles.

Dried Pigment Dried pigments are made by putting press cake into an oven, evaporating the water, and then grinding the dried product into a powder. Then the powder is dispersed into a vehicle to make the ink. Milling is a slow, expensive process, and if it is not carried out exactly right, the resulting ink has poor color strength and may contain grit.

Despite the slowness and expense of milling, dry pigment may still be preferable to many ink manufacturers because it can be dispersed into a number of vehicles. If, however, a pigment has been flushed into an alkyd varnish, it cannot be used to make a urethane ink or a water-based ink and poses some severe restrictions in making a UV ink.

Chips

Pigments for flexo, gravure, or screen inks are sometimes milled and dispersed into nitrocellulose, acrylic, polyamide, or other suitable resins. The dispersion is made into chips. The ink manufacturer dissolves the chips in the vehicle chosen to manufacture the ink.

Resinated Pigments

Treating the surface of the pigment with a suitable resin (rosin salt or a synthetic resin) before drying or flushing it makes it more easily dispersible by preventing agglomeration and making the pigment more easily wettable by the varnish.

Toners

Toner is a highly concentrated pigment (and/or dye), which is used to modify the hue or color strength of an ink. For example, alkali blue, the basic ingredient used in making reflex blue, is used to change the shade of black inks. The more alkali blue that a black ink contains, the more expensive it is.

Organic Pigments

The term **organic** means "derived from living organisms." In practice, most organic materials are made from petroleum, with lesser amounts made from coal, trees, and animal or vegetable fats and oils. By processing these materials, *synthetic organic* chemicals are made. All of them contain carbon and hydrogen, usually combined in one of a number of ways with oxygen, nitrogen, sulfur, and/or chlorine.

Materials containing only carbon and hydrogen are called *hydrocarbons*. Organic compounds are so numerous that a special branch of chemistry is devoted to them. This branch includes dyes and pigments, resins and plastics, and varnishes and solvents.

Black Pigments

Black pigments are extremely important, not only because black ink is used more than any other, but also because black pigments are the most permanent. They are practically unaffected by light, heat, acids, alkalies, and solvents.

Several black pigments have been used in printing inks: channel black, furnace black, thermal black, and lampblack. In the 1990s, furnace black is the only important black pigment. All of these blacks, called **carbon blacks,** consist mostly of elemental carbon, a small percentage of ash (mineral matter), and a somewhat higher percentage of volatile matter (depending on the method of manufacture). The volatile matter consists of chemicals containing carbon, hydrogen, and oxygen, usually produced during the incomplete combustion of petroleum feedstock.

Carbon blacks differ from each other in particle size, oil absorption, pH value, and volatile matter content. Structure, particle size, and surface chemistry are all significant in determining the behavior of carbon black pigments in ink. With carbon blacks, the finer the particle size, the higher the hiding power, or opacity; the longer the flow, the higher the viscosity; the higher the tack of the vehicle, the more work required to "wet out" the carbon in order to produce a suitable carbon dispersion. Because particle size and other properties differ with varied methods of producing the carbon blacks, ink properties depend upon the carbon black that is used.

Furnace blacks. Furnace blacks are the major group of carbon blacks. They are made by burning atomized mineral oil in bricklined furnaces with a carefully controlled supply of air. The products of combustion are cooled, and the pigment is collected with electronic precipitators or in bag filters.

The volatile matter content of furnace blacks is lower than that of either lampblacks or channel blacks. They have a bluer undertone and a higher pH value (7–10) than channel blacks (2–5). They also have less of a tendency to absorb driers and retard drying than do channel blacks or lampblacks.

Channel blacks. Channel blacks are made by burning natural gas with a limited supply of air. The flames impinge the surface of iron channels on which carbon black is deposited. The carbon black is then scraped off and collected.

Channel blacks made by the roller process (by impinging on iron rollers) have a smaller particle size, longer bodies (long flow), and bluer undertones than those made by impinging on channels. They also have high color strength and high gloss.

Despite their many excellent properties, channel blacks are rarely used now. The main reason being that they are no longer manufactured in the U.S. because of the environmental hazards they pose.

Lampblack. Lampblack is made by burning inexpensive unsaturated residues, such as creosote oil—a cheap by-product from the distillation of coal tar. The products of combustion are sent through a series of brick chambers where the carbon black particles settle out and are collected. Lampblacks have softer, grayer top tones and bluer undertones than the other carbon blacks, and they have a large particle size (the smaller the particle size, the blacker the masstone). They are soft, easily dispersed in ink vehicles, and often used in mixtures with channel and furnace blacks to reduce gloss and produce dull or soft finishes.

Colored Organic Pigments

Diarylide yellow. The diarylide yellows are readily ground on all types of ink mills and widely used in many types of inks. Pigment Yellow 12 is the most widely used. The diarylide yellows are highly transparent to red and green light, and they absorb blue light efficiently, giving them a hue error of almost zero and low grayness.

Spectral reflectance curve of diarylide yellow

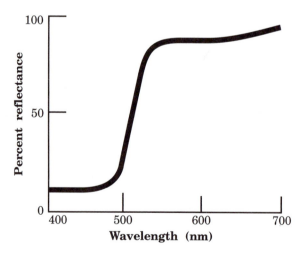

In addition to their good color, the diarylide yellows have a small particle size and are easily dispersed to produce an ink that flows well. Diarylide yellow has a higher tinctorial strength than Hansa yellow and is therefore preferred in printing ink.

Certain types of diarylide yellows are suitable for heat-resistant inks and where high tinctorial strength and brilliant tones are needed. They have poor to fair lightfastness.

If an opaque yellow is required, and chrome yellow is unsuitable, mixing diarylide yellow with titanium dioxide will yield an ink with both good color and opacity.

Hansa yellow. Pigment Yellow 1 is the most commonly used Hansa yellow. Hansa yellows are tinctorially less strong and less resistant to heat than diarylide yellows. They flow well because of their large particle size and have better lightfastness than the diarylide yellows.

Phthalocyanine blue. Phthalocyanine blue is the most important pigment for process blue (cyan) inks. It is resistant to a wide variety of chemicals and solvents and is very lightfast. However, in addition to its tendency to absorb blue and green light (making cyan colors characteristically too gray), phthalocyanine blue tends to give a bronze appearance to printed ink.

Phthalocyanine blue is available in alpha and beta forms. While the chemical properties are virtually identical, the alpha form converts to the beta form in certain solvents (including many used in gravure inks) and when heated. When the alpha form is converted to the beta form, a change in shade occurs, usually along with a loss in color or tint strength. The beta type, which is the form usually used in printing inks, is greener and cleaner than the alpha form. The beta form is used as the cyan in process printing. C.I. Pigment Blue 15 in the alpha crystalline form gives bright red-shade blues with high surface "bronze" whereas the beta-phthalocyanine blue, having a more compact structure, produces greener blues. These pigments have outstanding resistance to light, heat, acids, and alkalies.

Reflex or alkali blue. Although this pigment can retard the drying of oil-based systems, it still remains popular. Dispersed in lithographic varnish, heatset (or quickset system) reflex, or alkali, blue is used as a toner for carbon black inks, providing luster and masking the characteristic brownish tinge associated with carbon blacks. High tinctorial strength and good working properties are characteristic of alkali blue toners, which are used extensively in lithographic inks. Alkali blues tend to bleed in some applications, especially in alcohol.

Rubine. Azo red, the calcium salt of a complex organic acid, is the most commonly used rubine pigment. Its shade is yellower than a "true" magenta. It is used in litho and publication gravure applications. Care must be taken in its manufacture and usage to ensure suitable press performance in both offset and gravure. Lithographic printers should consult with their ink and dampening solution suppliers relative to potential bleed, scum, and other press performance problems.

Spectral reflectance curve of lithol rubine

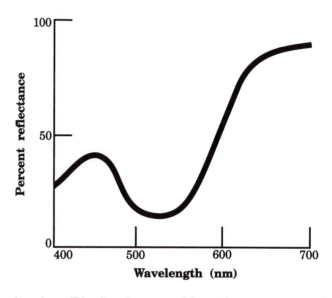

Rhodamine. Rhodamines are bluer (more magenta) than rubines and more expensive. Magenta inks can be prepared from either rubine or rhodamine or from blends, but rhodamine transmits more blue light and produces cleaner colors. Rhodamines tend to have poorer lightfastness and alkali resistance than rubines.

Red lake C. Red lake C is a common pigment in red inks. It is used in both paste and liquid inks. Chemically, red lake C is a close cousin of the rubine pigments, but its shade is considerably more yellow (that is, it absorbs blues). Use of red lake C in process inks lowers the cost, but it also limits the range of blues (the blue gamut) that can be printed.

Fluorescent Pigments

Fluorescent dyes are dispersed in inert, insoluble resins that are ground to a small size to produce fluorescent "pigments." These pigments are reputed to have poor lightfastness and poor flow when incorporated into ink. Recent advances have gone a long way toward overcoming these problems. Fluorescent pigments still work best in screen printing, although interesting effects have been achieved in other processes.

Inorganic Pigments

Many inorganic pigments are formed by precipitation—that is, by mixing solutions of chemicals that react to form the insoluble pigment, which then precipitates, or settles out. The pigment is separated by filtration, washed to remove soluble salts, and either flushed into a varnish or dried.

Inorganic pigments are often not pure chemical compounds. Many are complex mixtures, and the conditions of their formulation—namely, proportions, concentration, temperature, and pH value—must be carefully adjusted and controlled to get the desired color, brilliance, strength, texture, and fastness.

White Pigments

White inorganic pigments vary greatly in cost, specific gravity, particle size, and opacity when ground in ink vehicles.

Titanium dioxide. Titanium dioxide (TiO_2) is a brilliant white, opaque pigment. It is used extensively in package printing to provide an opaque white background for metal decorating and for printing on flexible packaging. These applications make titanium dioxide one of the most widely used of all ink pigments. Two forms of titanium dioxide, anatase and rutile (different crystal forms), are available. Rutile has a higher refractive index and greater opacity and is more abrasive. The anatase form is more stable and because of softer texture is preferred in gravure systems. Titanium dioxides have a high specific gravity and a particle size of about 0.2–0.3 micron. Inks made with them tend to pile on rollers, plates, and blankets unless properly formulated. They are the whitest and most opaque whites known.

Titanium dioxide is fairly readily ground on most types of ink mills. It is used in all printing processes, particularly where high opacity is required. It is most widely used in flexography and gravure because of its suitability for flexible packaging.

Because of its outstanding whiteness and opacity, titanium dioxide has largely replaced other white pigments in opaque tint bases and in white ink for metal decorating and decalcomanias. Since titanium dioxide tends to be wet by water preferentially, it is surface-treated to make it suitable for offset inks.

White materials such as clay and alumina hydrate are used as extenders in screen inks or letterpress inks. Extenders provide the suitable working properties that the colored pigments cannot provide.

Colored Inorganic Pigments

The principal advantage of colored inorganic pigments is that they are relatively inexpensive. Inks made from them have good opacity and lightfastness but can exhibit poor working properties and poor texture if not formulated properly.

Colored inorganic pigments that contain chrome or lead, such as chrome yellow and orange, molybdate orange, and cadmium yellow and red, have many desirable characteristics that made them useful in certain inks. However, because of the metals contained in these pigments, their use is limited by various environmental and health regulations.

Iron blue. The iron blues include milori, Prussian, Berlin, bronze, and Chinese blue. The various shades are obtained through different conditions of manufacture involving concentration, temperature, acidity, and method of oxidation. Use of iron blue pigments continues to decrease because phthalo blue performs better and is more economical.

Ultramarine blue. Ultramarine blue is a bright transparent blue produced from sulfur, silica, china clay, or carbon (rosin pitch or charcoal), and either soda ash or sulfate salts. This inexpensive pigment has excellent heat and alkali resistance.

Magnetic black. Black inks for magnetic ink character recognition (MICR) are made from iron oxide pigments, principally ferrosoferric oxide (Fe_3O_4) in a special crystalline form. Ferrosoferric oxide is used in magnetic inks designed primarily for the printing of bank checks. It causes the ink to be very short, therefore, when the ink is on press, it needs to be agitated constantly in order to keep it from backing away from the fountain roller.

Metallic Pigments

Metallic pigments of importance in the graphic arts are bronze (copper-zinc alloys ranging from 100% copper to 70% copper/30% zinc) and aluminum powders. Gold inks are made from bronze powders and silver inks from aluminum. Their manufacture is complex but generally requires the proper selection of metal alloys, careful control in ball milling to fine particle size, and appropriate use of fatty acids or sodium silicate for coating the particles. These metallic flakes are relatively coarse, their dispersibility in varnishes leaves much to be desired, and inks made from them frequently cause runnability problems.

Extenders

Thousands of tons of extenders are used every year in the manufacture of printing inks. These materials (kaolin clay, calcium carbonate, silica, and talc) alter the flow properties of inks (the lay of the ink, the ink holdout) and the color of the printed film. Extenders are required to have an extremely fine particle size (0.2–0.5 micron) to minimize abrasion and degradation of gloss. Clay content of gravure or flexo inks may be as high as 15–20% of total pigment, but lithographic inks seldom contain more than 2–4% extender. In screen inks, extenders increase the light scattering and whiteness of the printed film.

China clay. Hydrated aluminum silicate is known as china clay or kaolin. It is used in gravure and screen printing inks and as an extender for letterpress inks.

Calcium carbonate and magnesium carbonate. Calcium carbonate and magnesium carbonate (magnesia) are easily dispersed in ink vehicles, yielding transparent inks that have excellent working properties. However, inks containing these pigments tend to dry with a duller finish, and their dried films are more easily scuffed and marred.

Silica aerogel. Silica aerogel is almost pure silica. It is extremely porous and has a very high specific surface and oil absorption. When ground in varnishes, it is extremely transparent. It is used as a bodying agent in inks, where it increases their rate of setting and tends to prevent mottling and setoff.

Talc. Talc is a hydrated magnesium silicate. Micronized talc is used in printing inks to reduce gloss to give a velvet finish and reduce setoff and/or blocking.

5 Flow

The study of the flow of fluids (liquids or gases) is known as **rheology.** The printer is interested mostly in the flow of liquids (inks and varnishes) and controlling viscosity, body, and tack.

Viscosity

Varnish makers, ink manufacturers, and printers are referring to the **body** of an ink or varnish, when talking about its viscosity, consistency, length, or flow. Body is not precisely defined and, therefore, cannot be precisely measured. **Viscosity,** however, refers specifically to the degree to which an ink resists flow: the greater the resistance to flow, with a given applied force, the greater the viscosity. Inks are usually divided into two groups—paste inks and fluid, or liquid, inks—depending upon their viscosity. Paste inks are used for letterpress and lithography, and fluid inks are used for gravure and flexography. Because lithographic and letterpress inks have greater resistance to flow, they are (by definition) more viscous than gravure or flexographic inks.

Viscosity was first considered scientifically by Sir Isaac Newton in the seventeenth century. Newton reasoned that if a certain amount of force would turn a paddle or stirrer in a liquid, then doubling the force would double the speed of turning; that is, the velocity of flow was directly related to the force applied. (In scientific language, one says that the increase in the strain is directly proportional to the increase in the stress.) This is correct for many common materials such as water and gasoline, which are **shear-independent** or **Newtonian liquids.** At high rates of shear, flexographic and gravure inks are also shear-independent, or Newtonian, liquids.

Shear-Dependent Flow

For many materials, velocity of flow is not directly related to the force applied. These materials are referred to as **shear-dependent** or **non-Newtonian.** Paints and letterpress and lithographic inks fall into this group. Increasing the force on the stirrer or paddle does not increase the speed of the stirrer proportionally. The complexity of shear-dependent liquids is not yet well understood. The viscosity of a pseudoplastic-thixotropic liquid decreases as work is applied. When work stops, the liquid slowly regains its original viscosity. The two paths (the solid line and the dashed line) in the curve represent this increase and decrease.

Shear-independent,
or Newtonian,
liquid

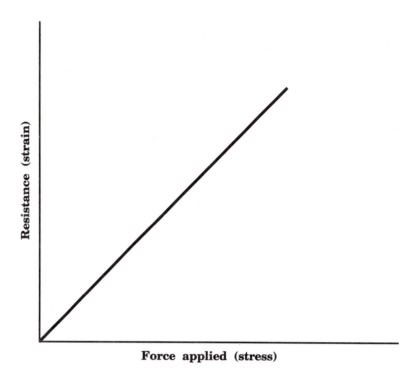

Shear-dependent
liquid, such as
flexo or gravure
ink

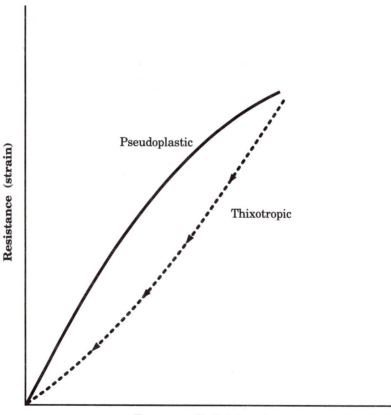

Yield Value

Some liquids will not flow at all until a certain finite amount of force is applied. These liquids are called **Bingham plastics,** and their behavior is illustrated in the accompanying illustration.

Bingham plastic

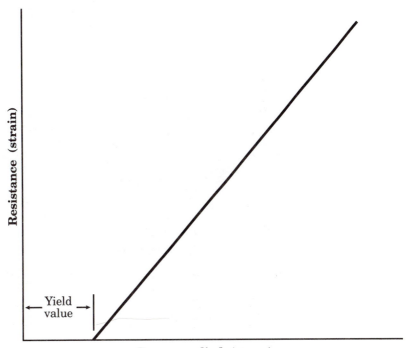

Force applied (stress)

For example, a pin or needle will float indefinitely in a can of lithographic ink. The lithographic ink behaves like a Bingham plastic. Similarly, mustard or mayonnaise in the bottom of a jar will remain there if the jar is inverted and allowed to sit on the shelf, and if a can from which most of the lithographic ink has been removed is inverted, it will not clean itself. On the other hand, flexo and gravure inks remaining in the bottom of a can will flow onto the cover if the can is inverted because they have low yield values.

Backing Away

Because paste inks have an appreciable yield value, they tend to hang up or "back away" from the roller in the ink fountain or in the ink duct. The ink does not flow under its own weight, and as the fountain roller removes all ink in contact with it, the printer observes ink in the fountain but none in contact with the roller.

A conical ink agitator applies force to the ink and keeps it flowing or prevents it from backing away.

Ink backing away
from fountain
roller

Ink

Fountain
roller

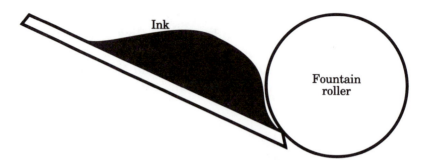

Conical ink
agitator

**Temperature
Coefficient of
Viscosity**

The viscosity of all liquids changes rapidly with the
temperature. The temperature coefficient of viscosity for
inks is about –3 or –4% per degree Fahrenheit (about –6%
per degree Celsius). This means that a temperature
increase of 15 °F (8 °C) will reduce the viscosity of the ink
by around 50%.

The high negative temperature coefficient of viscosity
has several implications for the printer. First off, if a
multicolor press is not warmed to working temperature
before printing starts, the ink flow and therefore the color
will change as the press warms up. Furthermore, should
the temperature on one of the units change, the flow of ink
will change on that unit, changing the color of the print.

In lithographic printing, if the press is cold when the
paper is fed to it, the highly viscous ink may adhere to the
paper and actually lift pieces of it from the sheet. This
problem is called picking. The press and the pressroom
must be up to working temperature before attempting to
feed paper into the press.

The temperature coefficient of viscosity also helps ink to set. The cooling of the ink as it passes from warm ink rollers onto cool plates and blankets greatly increases its viscosity. Most of the energy fed into a lithographic press is converted into heat on the ink rollers as they distribute and split the ink. The heat helps to make the ink fluid enough to flow readily, and it is an important reason for the complicated inking system found on lithographic presses. As the ink cools, going from roller to plate to blanket to paper, its viscosity increases rapidly. This contributes to the goal of every printer, the goal of "printing dry with wet ink."

Web offset presses increase the temperature of ink considerably. The great effect of temperature on viscosity (the high temperature coefficient of viscosity) is the reason that heatset web offset presses have chilled vibrators to cool the ink before it is placed on paper. Furthermore, in testing the viscosity of the ink in the laboratory, it is absolutely essential that the temperature of the measuring device be thermostated. If the temperature of the ink is not known, viscosity measurements are meaningless.

Thixotropy

When the temperature changes, the viscosity of all inks changes but paste inks also lose viscosity or body when they are stirred or otherwise sheared. This characteristic is called **thixotropy.** Ink taken directly out of the can will not flow as readily as ink that has been worked briefly on the slab. Ink that has been worked on the rollers has a lower viscosity than ink that is allowed to stand on paper. Thixotropy affects print sharpness and smoothness and also contributes to the problems printers encounter when they fail to mix additives thoroughly. Without working, the ink is stiff and the additives are not uniformly distributed.

Length

Length is the property of an ink that enables it to be stretched into a thread. The amount of stretching it withstands without breaking determines whether it is long or short. In general, long inks transfer and flow well, but they can also fly or mist. Short inks do not transfer as well and often cause printing problems. There is no relationship between the length of a material and its viscosity. For example, syrup has a low viscosity, but it is long. Mayonnaise, which is short, also has a relatively low

Comparison of a long ink *(left)* and a short ink

viscosity. Petroleum pitch, or asphalt, has a high viscosity and is long, but putty, which is also viscous, is very short.

The reader who is unfamiliar with these properties should try to use a knife to draw threads of mayonnaise and syrup from jars to get an idea of the differences in the length of the two materials.

The long materials, syrup and pitch, are relatively homogeneous; that is, they are uniform throughout. Mayonnaise, on the other hand, is a suspension of oil in water while putty is a suspension of solid in oil.

Imagine putting a piece of putty on the ink fountain or ink ductor roller of a press and then inching the press forward. The putty might crumble and drop off the press, but it would not uniformly coat the rollers and transfer down the press.

Offset inks must be long to prevent piling on the blanket. Screen inks must be short to prevent the formation of strings of ink when the screen is separated from the print. Altering or doctoring inks in the pressroom changes these important properties and causes printing problems.

Piling

Piling is the accumulation of materials on a blanket that originate from ink, paper, or a combination of both in sufficient quantity to affect print quality. It may occur in image or nonimage areas. Piling occurs in offset lithography. (See Chapter 12, "Lithographic Inks.") To begin with, offset inks are more highly pigmented than other inks because the lithographic process applies a thinner ink film than those applied by other printing processes. In the second place, offset inks are worked with a dampening solution, which emulsifies some of it. Therefore, there is already a potential problem when the ink on the blanket comes in contact with the paper. If small bits of dust (pigment or cellulose) accumulate in the ink, the ink may grow short, pile on the blanket, and fail to transfer.

The December 1988 edition of *Tappi Journal* features work on web offset ink piling by C. E. Coco and K. B. Cockerline. Their study shows that some highly absorbent papers may contribute to piling by absorbing vehicle from the ink very rapidly. In this case, the paper causes the ink to become short by removing the vehicle from the ink. If the ink fails to transfer to the paper, it cannot form a proper image on the sheet. Piling is therefore a paper/ink/press phenomenon. Although some papers will certainly resist piling more readily than others, ink and press conditions contribute to the problem or can be altered to overcome or reduce it. Unfortunately, printers sometimes cause piling by adding grease, vegetable shortening, corn starch, magnesia, or other solids to the ink.

Setting

Setting is the immediate buildup of viscosity after the ink is applied to the paper. It promotes ink trapping on multicolor lithographic presses. Setting is essential in sheetfed offset printing to avoid setoff (or "offsetting").

Setting is considered to be a physical phenomenon, while drying in sheetfed printing is considered to be a chemical occurrence. Drying and setting occur at the same time, and it is difficult to distinguish whether the increase in viscosity results from a physical phenomenon or from chemical behavior.

Viscosity increases rapidly when ink is chilled. The thixotropic nature of lithographic ink promotes setting, but ink manufacturers usually add a quickset varnish to further promote setting so that the ink printed on the second unit of a press will trap on the printed ink from the first unit of the press. Quickset varnish also enables the printer to print the backup side of the sheet sooner.

Since gravure and flexographic inks do not set, other techniques are required to trap a second-down ink. The most effective method used on flexo and gravure presses is to dry the film between each unit, e.g., drying the first-down ink before printing with a second ink.

Tack

Tack is the stickiness that can be observed when a printer taps out a thin film of ink on the slab or other flat surface. Offset inks should be as tacky or viscous as possible without damaging the paper when it is printed. The tackier the ink, the sharper and clearer it will print lines and halftones. It will also have less tendency to waterlog or break down and emulsify in the dampening solution. The tack of offset inks is always changed by emulsification with water or moisture picked up during printing. If inks are too tacky, they damage the paper and print poor solids; if inks are too soft, they print poor halftones and lines, and set off onto other sheets in the pile.

Scientists have been unable to define tack in terms of the physical properties of the ink. It is easier (and also more useful) to define tack as the resistance of the ink film to splitting, the force required to split a thin film, or the number obtained from the Inkometer or other tack-measuring device. (See Chapter 10, "Testing.")

In the late 1800s, a scientist named Stefan studied the forces required to split a thin film. He found that as the viscosity of the liquid between two plates was increased, the force required to split that film, or separate the plates, increased. Also, increasing the velocity of separation of the two plates increased the force, as did increasing the area between the two plates. He also showed that the force required to split a thin film was inversely proportional to the cube of the thickness of the film. These observations can be represented in an equation:

$$F = \frac{VSA}{t^3}$$

which shows that the force required to split a thin ink film on a press is related not only to the ink body itself, but also to the speed of the press and the area and thickness of the ink. The ink film of greatest interest to the printer is the one between the blanket and the paper, or, in the case of trapping, between the second-down ink and the print or the first-down ink.

As presented above, the "F" in Stefan's equation represents force, "V" is the viscosity of the ink, and "S" represents the speed of the press. (One should properly refer to the velocity of separation, but there is already one "V" in the equation, and printers refer to the speed rather than the velocity of the press.) "A" is the area of the film being split, and, as every printer knows, picking is most noticeable in solids where the ink film covers a large area between the blanket and the paper. Picking rarely occurs in halftone areas where there is reduced area of contact between paper and ink on the blanket.

The last term in Stefan's equation t^3, the cube of the thickness, may be hard to understand at first. We observe more picking or tail-end hook where the greatest amount of ink flows onto the press (in the solids area), and it might seem that increasing the amount of ink should make the paper stick more tightly to the blanket. This is not the case. Anyone who has worked with gauge blocks (pieces of flat metal) knows that they are harder to pull apart when a little oil is applied than when a thick film is applied. A thin film is harder to split than a thick film.

The fact that more ink flows down the inking train in the solids area than in the highlight area does not mean that the ink film between the paper and the blanket is thicker in the solids area, but, rather, that more ink is required to produce solids than to produce halftones, because the total inked area is greater. The thickness of ink on the plate should be uniform in all printing areas, halftones as well as solids.

Imagine an experiment in which we lay a sheet of coated or enamel paper on a two-foot-square polished chrome plate. It takes very little force to lift the paper. If we now apply a tiny dab of ink to that plate, roll it out, and press the paper against it, it will require somewhat more force to lift the paper and there will be very little ink on it. As we repeat the experiment, applying increasing amounts of ink to the plate, the force required to lift the sheet from the paper increases. The amount of ink on the paper increases gradually until there is enough ink to print a full solid. If we now increase the amount of ink beyond the amount required to print a full solid, the force required to lift the sheet from the block will decrease as increasing amounts of ink are added. (Note that because offset lithography uses the thinnest ink film of any printing process, it makes the greatest demand on the strength of the paper.) The experiment is summarized in the accompanying graph.

The graph shows that until there is enough ink on the plate to print a full solid, increasing the amount of ink does not increase ink film thickness on the paper; rather, it increases the area of contact. When enough ink has been

Effect of ink film
thickness on
splitting force

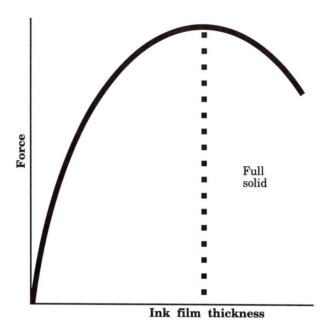

added to print a full solid, further additions of ink do not increase the area of contact between the ink film and the plate but, instead, increase the thickness of the film and reduce the amount of force required to remove the paper from the plate.

This experiment has been confirmed in the laboratory using a Vandercook proof press to roll a sheet of paper over a print wedge on the bed of the press.

Although Stefan's equation was derived from experiments with flat plates, pressroom experience shows that similar behavior occurs between cylinders; that is, between the inked blanket and the paper on the impression cylinder. If the paper cannot withstand the force of the splitting film, it picks, piles, lints, or curls. Decreasing the body or viscosity of ink, slowing the press, increasing ink film thickness, or reducing the area of the image all will improve paper performance.

If the printer adds a solvent or low-body varnish to reduce the body or viscosity of the ink, a thicker ink film is required in order to match color. The combined effect of lower viscosity and thicker film explains why reducing the body or viscosity of the ink is such a popular way to overcome picking, linting, and similar paper problems. Other materials are available that reduce the tack while having little effect on viscosity.

Even though reducing the body of the ink is widely recognized as a way to overcome paper problems, the fact that increasing ink film thickness also improves paper performance (blanket release) is not well known. Stefan's equation suggests that ink film thickness is even more important than body, speed, or area in affecting paper performance. If the equation, derived from flat-plate experiments, is applied exactly to rolls, then the following table shows the calculated decrease in required force due to the increase in ink film thickness. There appears to be no scientific evidence that the relationship derived for flat plates (as on a platen press) is directly applicable to a rotary press, but the general relationship should be roughly correct.

Increase in Ink Film Thickness (%)	Decrease in Release Force (%)
5	14
10	27
15	34

Estimated effect of ink film thickness on release force

Wet Trapping

Stefan's equation can also be used to improve the uniformity of printed color by improving trap.

Trap or trapping is the ability of a printed ink film to accept ink from the press. If the first-down ink has dried, the process is called dry trapping; if the print is still wet, it is called wet trapping. In printing on multicolor presses, wet trapping is of great importance in controlling color.

Numbers (% trap) are applied to the transfer of the second or succeeding ink films to the print. The calculation is based on the assumption that if the ink film transferred to the print has the same thickness as a film of the same ink printed on the paper, it will absorb the same amount of light. A densitometer is used to measure color density of the film on the paper and the film on the print.

The reflection densities of magenta, cyan, yellow, and their two-color overprints are used in the following formula to determine percent trapping.

$$\text{Percent trap} = \frac{D_{OP} - D_1}{D_2} \times 100$$

In this equation, D_1 is the reflection density of the first-down ink, D_2 is the reflection density of the second-down ink, and D_{OP} is the reflection density of the overprint.

These density readings are taken with the filter normally used for the second-down ink, e.g., a green filter when magenta is the second-down ink. This equation indicates *apparent trap,* because effects such as changes in gloss between single- and two-layer ink films could influence the percentage.

For example: if magenta over paper has an optical density of 1.30 using the green filter, magenta over cyan has an optical density of 1.44, and cyan over paper has an optical density of 0.39, the trap is 81%. If increasing the ink film thickness improves release (decreases the force required to split the film), one might expect that increasing the ink film thickness on the press, especially the blanket, would improve trapping.

This reasoning was used to solve a trapping problem reported to GATF. A printer, having trouble trapping cyan over yellow, called the ink manufacturer. The tack of the yellow (the first-down color) was 11 as measured on the Inkometer; the tack of the cyan was 9. The printer and ink manufacturer both expected good trapping performance. However, the thickness of the yellow ink measured 0.0003 in. (0.0076 mm) on the vibrator roller on the press, and the cyan measured 0.0002 in. (0.005 mm). When the ink manufacturer reduced the color strength of the cyan (forcing the printer to run a thicker ink film), the cyan trapped satisfactorily.

Because many printers recognize that ink film thickness affects trapping, some of them have established pressroom standards. The following pressroom standards illustrate control of ink and press to give excellent color control and trapping.

Sample specifications for printing inks	Color Sequence	Tack	Ink Film* Thickness	Ink Film** Thickness
	Black	17	0.20	5
	Cyan	16	0.28	7
	Magenta	15	0.32 or 0.34	8
	Yellow	13	0.40	10

Trap: M/C = 82%, Y/C = 90%, Y/M = 88%
* thousandths of an inch (mils)
** thousandths of a millimeter (microns or micrometers)

It should be readily apparent that the printer can use these specifications for only one color sequence. The above color sequence (K, C, M, Y) is very popular. There are many reasons for establishing and maintaining a single color sequence, and GATF recommends that each printer select and use only one sequence. If each ink applied on a multicolor lithographic press has succeedingly lower tack (as measured on the Inkometer), color uniformity of the printed product is improved. If the same set of inks is used in a different color sequence, color uniformity from print to print is impaired. Changing the printing sequence calls for a different set of inks.

It is inefficient and wasteful to change the color sequence after the press has been made ready. Every effort must be made to obtain inks that work the first time.

Ink manufacturers can provide unitack or monotack inks, which set very quickly so that they appear to increase in tack going from one press unit to the next, trapping the next ink. Many printers prefer these inks because one set of process colors can be used in any sequence. It is better to select one color sequence along with the pressroom standards that will make it work.

Ink Release/ Paper Release

Printers talk about "quick release" blankets almost as if the blanket picks up ink from the plate and releases it all to the paper. A brief inspection of an offset blanket on the press shows that this is not the case. The ink is not released from the surface of the blanket but splits at some point in the film.

Many years ago, Fetsko and Walker, of the National Printing Ink Research Institute, studied the transfer of ink from a metal plate to paper. Their study gives some insight as to what must be occurring not only between ink and paper but also between ink and blanket. In their study of ink transfer, Fetsko and Walker found that the percentage of ink transferred varied greatly with the amount of ink on the printing plate, as shown in the graph on page 54.

The curve may be explained as follows: when there is little ink on the plate, the paper contacts only a small amount of the ink and most of the ink remains on the plate. Even if the film splits 50/50 in the areas where the paper is in contact with the plate, the total transfer would be less than 50%. Increasing the ink film thickness increases paper coverage and ink transfer.

The peak on the curve is determined by paper smoothness. (On the press, speed and pressure also affect ink transfer.) Some of the ink is absorbed by the paper and

Effect of stock on
transfer of ink

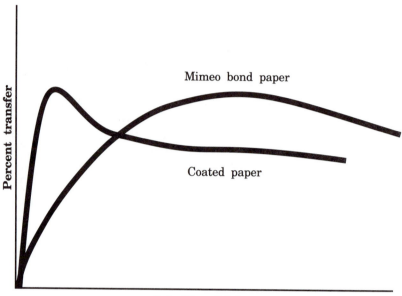

is, therefore, immobilized; it is not free to split. In fact, both roughness and absorbency of the paper seem to be responsible for immobilizing part of the film. It can be assumed that, except for the immobilized ink, the film splits approximately 50/50. The curve shows that as ink film thickness on the plate increases, the percent transferred gradually decreases.

In the splitting of an ink film between paper and blanket, there are two uneven surfaces. Some of the ink may be immobilized by the blanket as well as by the paper. Using the above reasoning, one might assume that a rough blanket would transfer less ink than a smooth blanket.

However, the printer must remember that a smooth blanket contacts paper at a much lower ink film thickness than a rough one does. Considering Stefan's equation, we would expect that a smooth blanket would provide a thinner film between paper and blanket, leading to poorer release and causing paper picking, embossing, curl, and tail-end hook. It may be assumed, then, that a rough blanket or a blanket that does not grow soft and smooth by absorbing ink oils will give better paper release.

This is the case demonstrated in GATF Research Progress Report 100. Of all the variables studied, only blanket smoothness had a major effect on paper release: rough blankets released paper far better than smooth blankets did. Hardness of the rubber or polymer material used to make the blanket had little effect.

From the above discussion, it appears that smooth blankets actually transfer ink more efficiently than rough blankets do. Yet, in demonstrations, rough blankets have released paper more easily than smooth ones have. Smooth blankets, however, do produce better solids and halftone dots than rough blankets do. The printer must make a suitable compromise between paper release and print quality.

Improvements in blanket manufacture in recent years have improved blanket performance, but the relationships still hold. Selection of the correct blanket is as important as selection of the proper ink if the printing is to proceed satisfactorily.

6 Vehicles

All printing inks consist of a colorant (almost always a pigment) and a **vehicle.** The vehicle is composed mostly of a varnish, which is a solvent plus resin and/or drying oil, along with waxes, driers, and other additives. The vehicle carries the pigment, controls the flow of the ink or varnish on the press, and, after drying, binds the pigment to the substrate. Vehicles also control the film properties of dried ink, such as gloss and rub resistance. (The word "varnish" [or "overprint varnish"] is also used to describe an unpigmented coating or film.)

Vehicles for lithographic and letterpress inks commonly contain low-molecular-weight resins and/or drying oils and nonpolar, hydrocarbon solvents. Inks made from these vehicles have a high viscosity and are called **paste inks.**

Vehicles for flexographic and gravure inks, which are referred to as **fluid** or **liquid inks,** have a low viscosity. These vehicles also contain a film-forming resin, a modifying resin, and a solvent. With flexographic inks, the solvent is often a polar solvent such as alcohol or water. The choice of solvent and resins depends on the substrate and end-use requirements of the ink.

Types of Vehicles

Oil-based ink vehicles are usually classified according to the primary feature of the vehicle. The major groups of oil ink vehicles include heatset vehicles, quickset vehicles, gloss vehicles, overprint varnishes, metal-decorating vehicles, infrared vehicles, and ultraviolet and electron-beam vehicles.

Heatset Vehicles

Heatset vehicles are composed of a resin dissolved in a petroleum distillate, heatset oil, or solvent. Although several manufacturers produce heatset oils, they are occasionally referred to as Magie Oils, a trademark of Magie Brothers Oil Company. Recent improvements in these very important vehicles have reduced the energy required for drying, the visible emissions from the dryer, and the amount of solvent required, while at the same time yielding films with higher gloss.

Quickset Vehicles

Quickset vehicles are particularly useful in sheetfed printing where high production speeds require fast setting, as with work-and-turn jobs. They are composed of two liquids that are marginally soluble in each other; for example, a high-viscosity oil and a solvent. When the ink is printed on an enamel stock, the low-viscosity solvent is quickly absorbed by the paper coating, leaving the high-viscosity oil on the surface. This means that quickset inks are less effective on uncoated than on coated paper, and they do not set at all when printed on nonabsorbent substrates.

Gloss Vehicles

Often called gloss varnishes, the high-viscosity gloss vehicles are made by adding a phenolic rosin-ester resin or a phenolic resin to a drying oil such as a linseed alkyd. In addition to gloss, these vehicles provide hard drying, good adhesion, or binding, and good resistance to dampening solution.

Overprint Varnishes

Linseed alkyds and other drying oils are typically reduced with heatset oil to provide a typical overprint varnish, but a wide variety of other formulations are also used. They are applied to wet or dry prints.

Water-based acrylic emulsions (water-based coatings) are often applied by wet trapping over offset litho ink films. They are applied either from the dampeners of the last printing unit or from applicators attached to the press. These aqueous varnishes dry very quickly, have little odor, resist yellowing on aging, and are compatible with most adhesives. These emulsions penetrate paperboard enough to allow glue to bond the surface, and there is no need for glue strips.

Water-based Gravure and Flexo Inks and Overprint Varnishes

Water-based gravure and flexo inks and overprint varnishes are based on polymer emulsions or alkali-soluble polymers that form a film on drying. If the alkali-soluble polymer is made soluble with ammonia, it becomes insoluble when the ammonia evaporates.

Metal-Decorating Vehicles

Polyester-based vehicles with excellent adhesion to aluminum, very fast heat drying (as rapid as 40 sec. at 375°F, or 180°C), high gloss, and resistance to abrasion and chemicals are replacing the older metal-decorating vehicles as two-piece aluminum cans replace the older three-piece lacquered metal cans.

Long-oil alkyds based on linseed, soybean, safflower, or dehydrated castor oil are often modified with other synthetic resins to make varnishes for flat-sheet tin.

Infrared (IR) Vehicles

Quickset vehicles have sometimes been referred to as infrared vehicles. Heat accelerates most physical and chemical processes, and infrared heaters added to the delivery end of the press accelerate the setting of quickset inks. IR heaters must be as carefully controlled as other functions of the press. A temperature of 105–110°F (42°C) in the printed stack gives fast setting and promotes drying. Higher temperatures tend to promote blocking (sticking the sheets together).

Ultraviolet (UV) and Electron-Beam (EB) Vehicles

Ultraviolet and electron-beam vehicles dry by radiation-induced polymerization. That is, the UV or EB radiation causes the vehicles to polymerize. Unlike the drying of sheetfed inks, this polymerization, or drying, is virtually instantaneous. EB is more powerful than UV, and UV inks and varnishes must contain a photoinitiator to catalyze the reaction. The vehicles are composed of oligomers, which are partially polymerized molecules such as epoxy acrylates, or acrylic esters, that are blended with a reactive acrylate diluent. The relatively high cost of these materials has limited their use to specialty products for which the cost is justified: album and textbook covers, folding cartons, plastics, foils, and metal decorations.

Components

The types of components found in printing ink vehicles are listed in the table on page 60.

Drying Oils

The important drying oils are linseed, tung (chinawood), dehydrated castor, safflower, soybean, tall, and oiticica. Unmodified vegetable oils are rarely used in printing inks. They are almost always modified chemically or by heat or bodying before they are used.

   Linseed oil, a pale, straw-colored oil, is produced from the seeds of the flax plant. After extraction, impurities are removed either by acid or by alkaline refining. The product is washed, dried, and filtered to yield the refined oil.

   In the production of *boiled linseed oil,* heat [475–575°F (245–300°C)] is used to thicken or increase the viscosity of the oil. The process is commonly referred to as **heat-bodying** and the result is a litho varnish, frequently called a heat-bodied linseed oil (HBLO), or "stand oil." During heating, molecular units link together to form progressively more complex structures of increased

Components of vehicles for printing inks

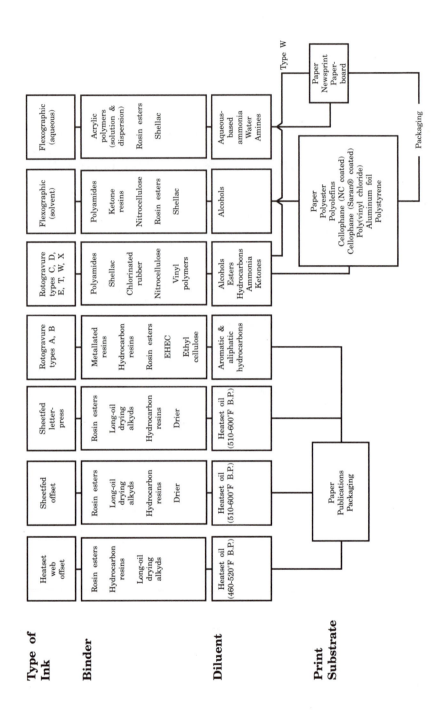

Type of Ink	Heatset web offset	Sheetfed offset	Sheetfed letter-press	Rotogravure types A, B	Rotogravure types C, D, E, T, W, X	Flexographic (solvent)	Flexographic (aqueous)
Binder	Rosin esters, Hydrocarbon resins, Long-oil drying alkyds	Rosin esters, Long-oil drying alkyds, Hydrocarbon resins, Drier	Rosin esters, Long-oil drying alkyds, Hydrocarbon resins, Drier	Metallated resins, Hydrocarbon resins, Rosin esters, EHEC, Ethyl cellulose	Polyamides, Shellac, Chlorinated rubber, Nitrocellulose, Vinyl polymers	Polyamides, Ketone resins, Nitrocellulose, Rosin esters, Shellac	Acrylic polymers (solution & dispersion), Rosin esters, Shellac
Diluent	Heatset oil (460-520°F B.P.)	Heatset oil (510-600°F B.P.)	Heatset oil (510-600°F B.P.)	Aromatic & aliphatic hydrocarbons	Alcohols, Esters, Hydrocarbons, Ammonia, Ketones	Alcohols	Aqueous-based ammonia, Water, Amines

Print Substrate

Paper, Publications, Packaging

Paper, Polyester, Polyolefins, Cellophane (NC coated), Cellophane (Saran® coated), Poly(vinyl chloride), Aluminum foil, Polystyrene

Paper, Newsprint, Paper-board

Type W

Packaging

molecular size. Viscosity increases with the increase in molecular size: the longer the oil is heated, the greater the molecular size and the higher the viscosity of the oil. The time required to obtain a given viscosity depends on the specific composition of the oil, the temperature, and the absence or presence of air and a catalyst.

Litho varnishes are graded from #00000 (the thinnest or least viscous) through #0 and consecutively #1 through #8 (the stiffest or most viscous), sometimes including #9 and #10 grades. However, no general standards of viscosity have been adopted and each varnish manufacturer has individual standards based on viscosity measurements, specific gravity, and iodine number.

Another means of controlling the viscosity and other properties of oil during its production has been to blow large volumes of air through the oil at a relatively low temperature, 210–280 °F (100–140 °C). The product of this exothermic reaction (heat is given off) is a **blown oil.** Blown oils have a higher oxygen content than litho varnishes have. In litho varnish production, chemical oxidants are used instead of blowing to obtain the desired properties.

Blown oils generally provide better pigment wetting and dispersion, flow properties, and drying rate than litho varnishes, but they are darker colored and have a stronger odor. They also tend to liver and are considerably less stable in storage than litho varnishes are.

Linseed and other drying oils dry as a result of polymerization of unsaturated carbon-to-carbon bonds after initiation by oxygen. Because the structure contains an allylic group, polymerization takes several hours.

There are methods (distillation and solvent extraction) to enrich the presence of reactive molecules. Modified linseed oil is sufficiently reactive for most printing ink applications, but other oils can be made more useful by enrichment or by chemical modification.

Tung oil, also called "chinawood oil," is obtained from the seeds of tung trees. Originally grown only in China, the tung tree is now also grown in the southern U.S. and in Japan. The oil dries very rapidly (faster than linseed oil), and its molecules contain more unsaturated carbon-to-carbon bonds than linseed oil molecules contain.

The speed of polymerization tends to cause gelling upon heating. The ink film may be frosted, flattened, or crystallized (gas crazed) upon drying. Varnishes based on tung oil dry very quickly and become very hard, creating dry trap problems (it is hard to print over the dried film). Printers say the ink has "crystallized." The ink manufacturer takes these factors into account to formulate an ink that will improve the benefits and control the problems.

Castor oil originates from the castor bean, a plant grown primarily in Brazil, East Africa, and India. It has the highest viscosity of any vegetable oil. It is not a drying oil, but when heated, water is split out of the molecule, producing **dehydrated castor oil** (DCO). DCO, resembling tung oil, dries more rapidly than linseed oil. Its primary use is in vehicles for white lithographic metal-decorating inks, where its superior nonyellowing characteristics are important.

Oiticica oil is a natural oil derived from the fruit kernels of a wild tree in Brazil. Although darker in color and higher in viscosity and acid number, it resembles tung oil in drying speed and can be used as a substitute.

Tall oil, which is obtained from the kraft pulping of wood chips and thus contains natural wood oils, can be modified to produce a drying oil. Tall oil is separated from the fatty portion of black liquor (from pulping) and reacted with resins to yield tall-oil alkyds or tall-oil phenolics, which are inexpensive and valuable ingredients of inks.

Soybean and safflower oils are used as alkyd derivatives in making overprint varnishes and white or light-colored lithographic metal-decorating inks because of their good nonyellowing characteristics.

Primary oil ink vehicles. Primary oil ink vehicles are drying oils such as linseed oil that have been chemically modified with various alkyds, phenolics, urethanes, or modified rosin esters, or other synthetic resins, depending on the properties desired in the ink or varnish. To produce a linseed isophthalic alkyd, a dibasic acid, such as isophthalic acid, is reacted with glycerine and the resulting material is reacted with linseed oil.

Modified Drying Oils

Alkyd resin vehicles. If the vegetable oil contains a small amount of synthetic resin, such as an alkyd resin, the product is called a long-oil synthetic, because it has a relatively large percentage of oil. If the oil is reacted with a large amount of alkyd resin, then it is called a short-oil synthetic because it has a relatively small percentage of oil. Medium-oil synthetics contain approximately equal amounts of oil and alkyd resin. Not surprisingly, short-oil synthetics more closely resemble the alkyd resin in properties and long-oil synthetics more closely resemble the oil in properties.

Alkyd resin vehicles dry faster and require less drier and less oxygen than vehicles made from unmodified drying oils. Alkyd resin vehicles have good wetting properties for pigments, and often produce inks with higher pigment content and less tack than linseed varnishes can, with or without additives or solvents. Alkyd resin vehicles are used extensively in overprint and gloss varnishes.

Alkyds. Alkyd resins improve pigment wetting, film toughness, and adhesion. Alkyds are important resins used in paints, varnishes, and printing inks. The major alkyds used in printing ink today are based on isophthalic acid. They are produced from linseed, tung, tall, dehydrated castor, and soya bean oils. Oil-modified alkyd resins are primarily used in letterpress and lithographic ink. Non-drying-oil alkyds are excellent plasticizers for nitrocellulose in gravure inks.

Phenolics. Phenolic resins of interest to the printer are made from formaldehyde and phenol derivatives called alkylphenols. Pure phenolic resins are reacted with drying oils to produce litho varnishes with excellent wetting characteristics and stability, which yield tough, glossy films with good adhesion and alkali resistance.

Alkylphenol resins are compatible with long-oil alkyds, drying oils, and modified rosin esters. Alkylphenol resins are commonly used in high-gloss vehicles and inks. Alcohol-soluble, pure phenolics are used in flexo inks and as modifiers for top lacquers.

Urethanes. Urethanes, chemical cousins of the polyamides, can be combined with drying oils, as can the alkyds and phenolics, to yield varnishes of use in litho and letterpress ink. Incorporation of a urethane normally improves adhesion and chemical resistance.

Nondrying Oils

Mineral oils, obtained from petroleum, are classified as nondrying oils because they do not dry by oxidation. However, they do dry by absorption or evaporation and are used in inks for newspaper and other web printing.

Resins

Low- or mid-molecular-weight organic polymers, the sole binders in many web inks, are often incorporated into sheetfed inks. They vary widely in composition: some are of natural origin, some are synthetic. High-molecular-weight resins lack solubility in solvents used for printing inks.

Resins are often referred to as thermosetting (those that react to form a rigid product that will not melt) or thermoplastic (those that soften or melt when heated).

Hydrocarbons. Resins derived from hydrocarbons are valuable in formulation of high-performance varnishes for litho sheetfed and web inks. Hydrocarbon resins, by-products of petrochemical processing, are used to replace part or all of the rosin derivatives in gravure and web offset inks. They are easily soluble in ink solvents and very compatible with other oil ink resins. They have a broad range of melting points and compatibilities. The low-melting grades, used as extenders in gravure, lithographic, and letterpress ink vehicles, exhibit poor solvent release and gloss. Higher-melting resins—285°F (140°C)—have excellent solvent release, promote quick setting, and build gloss.

Low-cost, black gravure inks incorporate asphalt and pitch, which occur naturally or as a residue from the distillation of petroleum or coal. Gilsonite is a natural pitch or bitumen that promotes pigment dispersion. Its dark color limits its use to black inks.

Natural Resins

The high price and shortage of shellac, which originates as the natural secretion of an insect (Laccifer bacca), have resulted in its replacement with acrylic emulsions in water-based flexo inks. The insect secretion, or lac, is refined by dissolving it in alcohol and decanting or filtering it.

Modified Natural Resins

Cellulosics. Cellulose is a widely distributed and chemically reactive natural material. It is used in many ways that are of interest to ink manufacturers and printers. In its fibrous form, it is the most important ingredient of paper. The chief industrial sources of cellulose are trees and cotton. The polymer occurs in fibers and is insoluble in most solvents.

Cellulose can be reacted with a variety of chemicals to convert it to products that are soluble in many ordinary solvents. Cellulose derivatives form films that are useful not only in printing inks, but also in a wide variety of packaging and other industrial applications.

Cellulose polymers—thermoplastic materials with high melting points and high molecular weights—give strong coherent films by solvent evaporation. They show excellent solvent release, good flexibility, compatibility with other resins, and good adhesive properties.

When reacted with nitric acid, cellulose forms an ester, cellulose nitrate (usually called nitrocellulose) that is used for flexographic and gravure inks and coatings. It gives tough, brilliant films, and has excellent pigment-wetting, solvent-release, and heat-resistance properties. Nitrocellulose is flammable at high temperatures, but at temperatures below this point, its resistance to softening is higher than that of other films, such as polyethylene or polypropylene. High levels of nitration yield a product that is soluble in esters and ketones and suitable for gravure inks. Lower nitration yields a product with greater alcohol solubility that is suitable for flexography.

When reacted with acetic anhydride, cellulose forms cellulose acetate, a product that is soluble in esters and ketones. Cellulose acetate films are less flammable than cellulose nitrate films.

In the presence of aqueous caustic, cellulose can be reacted with ethyl chloride to make ethyl cellulose, with chloroacetic acid to make carboxymethyl cellulose, or with ethylene oxide to make hydroxyethyl cellulose.

Cellulose nitrate and ethyl cellulose, in combination with other resins, are used in gravure and flexographic inks for printing on paper, aluminum foil, and plastic surfaces. Ethyl hydroxyethyl cellulose is used as a film former in gravure and screen printing inks, and sodium carboxymethyl cellulose is used as a thickening agent for water-based inks.

Cellulosic resins used in printing inks	Abbreviation	Name
	CMC	Carboxymethyl cellulose
	HEC	Hydroxyethyl cellulose
	EHEC	Ethyl hydroxyethyl cellulose
	CMHEC	Carboxymethyl hydroxyethyl cellulose

Rosins. Rosin is a natural resin, but it is modified chemically for use in inks. There are three types of rosin, classified according to source: gum, wood, and tall oil. Gum rosin is obtained from the sap of tapped trees. Wood rosin is derived from the stumps of pine trees. The stumps are mechanically disintegrated, and the rosin is extracted with hot naphtha. Tall-oil rosin is a by-product of the wood pulping operation in the manufacture of paper. It is isolated by acidification of the black liquor soap that remains after kraft paper pulping and fractional distillation of the resulting crude tall oil.

Rosin consists of rosin acids (about 90%) and neutral material. Before it is used in printing ink, the rosin is chemically modified. It may be reacted with penta-erythritol, maleic anhydride, and various phenols to produce a range of rosin ester resins with varying melting points and solubilities.

In inks for publication gravure, both calcium and zinc/calcium resinates are widely employed. These resinates are usually called limed rosin or zincated rosin. Reacting rosin with zinc or lime raises the melting point, improves solvent release, reduces free acidity (thereby providing for less pigment reactivity), increases brittleness, and prevents crystallization.

The glycerol ester of rosin, estergum, is produced by reacting glycerol with tall-oil rosin. The product, with a ball-and-ring softening point of around 176°F (80°C), is used in modified linseed oil varnishes for preparation of letterpress and lithographic inks. Estergum is no longer used in high-performance grades of printing inks, but can be used in low-cost publication inks printed by web offset, rotogravure, or rotary letterpress.

The pentaerythritol ester of tall-oil rosin has a higher ball-and-ring softening point, around 212°F (100°C), and a faster solvent release than the glycerol ester. Penta-erythritol esters of polymerized rosin have still higher softening points, 265–285°F (130–140°C), are extremely stable in solution, and are used in heatset lithographic inks.

The pentaerythritol ester of dimerized rosin (Pentalyn K®) is a widely used rosin. It has a high softening point, 350–365°F (175–185°C), is soluble in aliphatic hydro-carbons, resists hydrolysis and emulsification, wets pigments well, and is stable with reactive pigments. Penta esters of dimerized rosin are used in flushing and in heatset letterpress, offset, quickset, and gravure inks.

Rosin-modified phenolic resins. Modifying rosin esters with various phenols produces harder resins with higher melting points and reduced solubility. They are used in oil ink vehicles for quick setting, good solvent release, rub resistance, high gloss, and resistance to litho dampening solutions. Various rosin-modified phenolics are used in heatset vehicles and gels, sheetfed quickset vehicles and gels, and gloss varnishes, either alone or in combination with other resins.

Rosin-modified maleic and fumaric resins. These are useful in water-based flexo inks and in water-reducible letterpress inks. When reacted with glycols they are used in gravure and flexo inks, sometimes in combination with nitrocellulose or polyamides. Alcohol-soluble (glycol-soluble) maleic resins are used in both solvent and aqueous flexographic ink systems. Oil-soluble maleic resins are commonly used in gravure, heatset offset, and quickset inks, as well as in overprint varnishes.

Rosin-modified maleic resins are available with a range of properties, which are dependent upon the type and amounts of rosin, dibasic acid, and polyol. Gum rosin has a higher viscosity and melting point than wood rosin and tall-oil rosin; however, polymerized rosin has the highest melting point. A high ratio of dibasic acid to polyol usually increases hardness and viscosity but reduces oil solubility.

Cyclized rubber. The treatment of solid natural rubber with strong acids results in loss of elasticity and the formation of a resinous substance called cyclized rubber.

Properties of the final product depend upon the ingredients used, the reaction conditions, and the degree of reaction. Rubber treated with sulfuric acid yields a low-molecular-weight product, the type preferred for printing inks. Because of its solubility in aliphatic and aromatic hydrocarbons, good pigment wetting characteristics, high gloss, and excellent scuff resistance, low-molecular-weight cyclized rubber is useful as a modifier for resinates and varnish-maleics in gravure type A and B inks for magazine printing. The relatively high cost of this resin limits its use as a sole binder.

Rubber-based ink is a popular type of offset duplicator ink, where cyclized rubber contributes very fast setting and excellent "stay-open" properties on the press.

Synthetic Resins

Vinyls. Vinyl resins are commonly made into films that may be printed by a variety of processes. Vinyl resins are sometimes used in inks for screen printing, notably in plastisol inks. Poly(vinyl chloride) is a hard, brittle material that cracks readily. To make it tougher and more pliable, it must be plasticized. This can be done by incorporating vinyl acetate into the polymer, by adding a softening agent (plasticizer) to the dry resin, or both.

A **plastisol** is a suspension of powdered resin in a plasticizer. The hard, dry powdered resin is not very soluble in phthalates or other plasticizers, but it dissolves easily when heated. Adding a dye or a pigment converts it into an ink, which can be printed (usually on textiles) and converted to a tough film on heating.

Vinyl copolymers, such as copolymers of vinyl chloride and vinyl acetate, are soluble in many solvents and are used to make inks for screen printing.

Styrene-maleics. Copolymers of styrene and maleic acid are soluble in alkali and in ammonia. They have been used to make low-cost, water-soluble flexo inks. The dry films lack gloss and rub resistance.

Ketones. Ketone-formaldehyde condensates are light-colored resins used as modifiers of nitrocellulose, ethyl cellulose, acrylic, and polyamide resins to improve gloss and adhesion in flexo inks. These low-viscosity resins possess excellent solubility in alcohols and in ketone and ester solvents.

Polyamides. Polyamides are used in high-quality flexo and gravure packaging inks, primarily for printing untreated polyethylene and other flexible packaging films. They can be made soluble in alcohol. Alcohol-soluble polyamide resins are used to make flexo and gravure inks that have excellent adhesion to polyethylene, foil, and other films. They have high gloss, resistance to fats and oils, good solvent release, and good compatibility with nitrocellulose. Polyamides that are soluble in an alcohol and hydrocarbon mixture are referred to as cosolvent polyamides.

Acrylics. Acrylic resins, polymers or copolymers of acrylic or methacrylic esters, are widely used in industry. Acrylics are used in a variety of solvents as binder components in flexographic, gravure, and screen printing inks.

Acrylic resins have excellent adhesion to most packaging films and foils, outstanding resistance to discoloration by ultraviolet light and heat, and resistance to grease, oil, and water. These resins can be formulated for solubility in almost any solvent system used in modern printing inks. Compatibility with other commonly used binders, such as nitrocellulose, cellulose acetate butyrate, chlorinated rubber, vinyl resin, and many rosin derivatives, provides broad latitude in formulating to meet specific performance requirements.

Ethylene-acrylic acid. Copolymers of ethylene and acrylic acid dissolved in aqueous organic bases or ammonia form water-based flexo or gravure inks that dry to make hard, glossy prints with excellent adhesion.

Epoxies. Epoxy resins, which are soluble in ketones, esters, and alcohols, are compatible with certain polyamide, formaldehyde, phenolic, and vinyl resins. They are usually incompatible with alkyd resins, cellulose and rosin derivatives, and vegetable oils. Epoxy resins produce films that are tough, durable, flexible, and very resistant to abrasion, moisture, and chemicals. Epoxy esters based on drying and nondrying fatty acids can be modified with styrene or rosin. Epoxy esters are used primarily in offset metal decorating inks (for collapsible tubes), varnishes, and special screen process inks.

Poly(vinyl butyral). This white powder is soluble in most common solvents. It provides excellent adhesion to glass, metal, and plastics and is a major component of electrostatic toners.

Alkylated urea- and melamine-formaldehyde. When formulated with alkyd or polyester resins, these resins are used in metal decorating inks, where the baking of the product converts them into tough, inert films with good adhesion to metal. Both resins are used in high-gloss lacquers and flexographic and gravure inks where chemical inertness of the finished film is needed.

Coumarone-indene resins. The coking process in the steel industry yields fractions rich in indene. Resins polymerized from these fractions are called coumarone-indene, or coal tar, resins. They show excellent leafing properties in aluminum and bronze powder inks. Although inks made with coumarone-indene resins are still in use, these constituents of gravure, flexographic, and oil-based inks have been largely replaced with hydrocarbon resins, which are lower priced, and show better color retention.

Terpenes. Terpene resins, prepared from turpentine, are pale yellow, neutral, and chemically inert. They show good color retention upon exposure to heat. They are particularly suited to the formulation of high-gloss overprint varnishes for letterpress and lithography. Terpenes that are water-white from polymerization of pure monomers or from hydrogenation of cycloaliphatic resins have recently gained increasing use in such overprint applications.

Solvents

Solvents dissolve oils, resins, and additives to produce varnishes that carry the pigment. Properties of some common solvents are listed in the accompanying table.

Properties of solvents used in inks

Solvent	Density (g/mL @ 20 °C)	Specific Gravity (lb./gal.)	Boiling Point °F	Boiling Point °C	Flash Point °F	Flash Point °C
n-Heptane	0.684	5.70	208	98	20	−7
Xylol (1,4)	0.861	7.18	280	138	82	28
Toluol	0.866	7.22	232	111	41	5
Cellosolve*	0.930	7.75	275	135	120	50
Methyl Cellosolve*	0.965	8.05	257	125	120	50
Butyl Cellosolve*	0.901	7.53	340	171	165	74
Ethyl acetate	0.901	7.52	171	77	24	−4
Isopropyl acetate	0.872	7.27	192	89	35	1
n-Butyl acetate	0.882	7.36	259	126	76	24
Ethyl alcohol	0.789	6.58	172	78	54	12
Isopropyl alcohol (anh)	0.786	6.55	180	82	53	12
Isopropyl alcohol (91%)	0.818	6.84	176	80	61	16
Methyl ethyl ketone	0.805	6.71	176	80	16	−10
Methyl isobutyl ketone	0.798	6.65	243	117	60	15
Cyclohexanone	0.948	7.91	313	156	129	54
Isophorone	0.923	7.70	421	216	200	93

*Cellosolve is the registered trademark of Union Carbide Corp.

The old rule "like dissolves like" is a useful first approximation. Because of the importance of solvency, scientists have studied it extensively and have come up with sophisticated measurements of solvency. The "solubility parameter" has proven very useful in predicting solubility. Polar solvents, such as alcohol, ethers, ketones, and esters, are useful for dissolving polar resins, such as shellac, cellulose esters, phenolics, and alkyds. The hydrocarbons, which are nonpolar, are suitable for dissolving such compounds as drying oils and rosin-modified phenolics and maleics. The different classes of hydrocarbons vary in their degree of solvency.

Diluents and Thinners

Thinners or diluents are added to an ink or varnish to reduce its pigment concentration or resin concentration. A **thinner** is a solvent for the vehicle, and a **diluent** is a nonsolvent or a poor solvent. Thus, an ester or a ketone would be thinner for a rotogravure ink, while alcohol would be called a diluent.

Diluents may be added to resin solutions to change the rate of evaporation or to reduce the cost. A thinner may also be added to reduce the viscosity of the ink mixture.

In addition to solvency power, the boiling point or range and evaporation rate of the solvent are important to ink drying and print quality.

In offset and letterpress printing, ink is distributed over many rollers before it reaches the printing plate. If the solvent is too volatile, the loss of solvent will cause a rapid increase in ink viscosity and tack, which can cause picking or tearing of the paper being printed. If the solvent is not sufficiently volatile, it creates drying problems in any printing process that depends on evaporation for drying.

Safety

Most solvents are flammable and explosive, and they are often toxic. In addition, their odor may be objectionable. Solvent manufacturers or distributors should be consulted about potential hazards before using them. It is important to know some of the fundamental factors involved.

Commonly used solvents, with the exception of some chlorinated compounds, are flammable. The degree of flammability of a solvent is generally indicated by its **flash point,** the lowest temperature at which the substance gives off vapor that will ignite. The lower the flash point, the greater the hazard.

The **ignition point** is the temperature at which the vapor given off by the liquid continues to burn after ignition. It is higher than the flash point and is determined by such factors as heat of combustion, the supply of oxygen, the rate of heat diffusion, and the type of surface involved.

Flammable solvents, when combined with the air, form explosive mixtures under certain conditions of temperature and pressure. The lower explosive limit of a solvent is the minimum concentration of its vapor in air that can be ignited by a spark or a flame and burn by self-propagation. The upper explosive limit is the maximum concentration of its vapor in air that can be ignited by a spark or flame and burn by self-propagation. At a temperature of 68°F (20°C), the lower explosive limit of acetone, for example, is 2.89% by volume, while the upper limit is 12.95%.

The range between the low and high levels (or minimum and maximum volumetric ratios of solvent vapor and air) is called the **explosive range.** Below the lower limit, the dispersion of solvent in air is too limited to allow combustion to spread, even though one segment is exposed to ignition. Above the upper limit, the solvent is so concentrated that there is insufficient oxygen for combustion.

Accordingly, safety precautions should be carefully followed. Solvents should be stored only in approved containers and handled in a specified manner.

Some solvents exhibit acute narcotic effects that dissipate shortly after exposure, and others are characterized by cumulative toxic effects. For example, the U.S. Bureau of Mines reports that the body can tolerate occasional exposure to small amounts of methanol, but the safe exposure limits for commonly used ester solvents, ketones, and petroleum naphtha are not known. Material Safety Data Sheets (MSDS) on these materials should be read before handling them.

Hydrocarbons   Two classes of hydrocarbons are of prime interest to the printer: the aliphatic hydrocarbons, like those dearomatized solvents used in heatset, and aromatic hydrocarbons, as typified by toluene and xylene. Kersosene is a petroleum product defined by boiling range. It contains a mixture of ally hydrocarbons. Kerosene, however, also contains carcinogenic materials while heatset oils are treated to remove carcinogens. Some common aliphatic hydrocarbon solvents are listed in the accompanying table.

Petroleum solvents

	Boiling Range		Flash Point*		Evaporation Time
	°F	°C	°F	°C	(sec., 100°)
Varsol 3†	313-347	156-175	105	40	2,290
Lactol‡ spirits	206-214	97-101	20	−7	—
VM&P naphtha	244-282	118-139	44	7	350
Mineral spirits	317-390	158-199	112	44	5,420
High-flash mineral spirits	330-400	165-204	140	60	—
Stoddard solvent	340-385	171-196	105	40	—
Kerosene	356-570	180-300	100	38	—
Hexane	149-157	65- 69	−10	−23	90

*TCC, ASTM D 56
†Trademark Exxon Corporation
‡Trademark Amsco

Sources: *Industrial Solvents Handbook,* 2nd ed., Noyes Data Corp.
Condensed Chemical Dictionary, 10th ed., Van Nostrand Reinhold, revised by Gessner G. Hawley

Aromatic hydrocarbons are more powerful in dissolving resins than aliphatic hydrocarbons are. Toluene is used extensively in gravure inks. Xylene is used in gravure proofing inks because it evaporates more slowly than toluene. Toluene is identical to toluol, and xylene is identical to xylol, except that, for shipping purposes, toluene and xylene are classified as chemicals and carry a higher freight rate than toluol and xylol, which are classified as solvents. In the trade, these substances are always called toluol and xylol.

Mineral spirits, naphtha, Stoddard solvent, and kerosene consist principally of aliphatic hydrocarbons. Solvents of this type are used in web inks for lithographic, letterpress, and gravure printing.

In flexographic inks, aliphatic hydrocarbons are primarily used as cosolvents in polyamide formulations or as inexpensive diluents. Aromatic hydrocarbons are used only sparingly in flexographic inks because they swell rubber and plastics.

Heatset Oils These carefully fractionated, narrow-boiling, hydrocarbon oils are solvents for most of the resins found in inks. They are used in most types of lithographic and letterpress inks. The choice of solvent with the proper boiling range depends on the type of ink. Heatset oils are produced by several manufacturers and differ slightly from one company to the next. Printers must consult individual manufacturers to obtain the specific properties of the heatset oils that they produce.

The boiling points of aliphatic hydrocarbon solvents for heatset inks vary from about 400 °F to 600 °F (200 °C to 300 °C), but each designated fraction boils over a much narrower range. Choice of heatset oil depends principally upon the type of resin used and the printing conditions for which the ink is intended. The lower the boiling point, the lower the temperature required to dry the print. Selection of the solvent depends not only on the boiling point, but also on the characteristics of the dryer and the press, press speed, ink coverage, and paper stock. Special heatset rubber rollers are needed to minimize solvent absorption.

Quickset sheetfed inks for letterpress and lithography contain aliphatic hydrocarbon solvents with initial boiling points of approximately 510 °F (265 °C) and higher.

Alcohols

Alcohols are extensively used in flexographic and gravure inks and in some screen printing formulations. Their solvency power, evaporation range, and mild effect on rubber make them important flexographic solvents.

Alcohol solvents include methyl (methanol), ethyl (ethanol), normal propyl (n-propanol), and isopropyl (isopropanol) alcohols. All of these materials have a degree of toxicity and should be handled with respect.

Among the alcohols, methyl alcohol (methanol, or wood alcohol) evaporates the fastest and is a very good solvent for dyes. Ethyl alcohol (ethanol, or grain alcohol) must be denatured to render it unfit for human consumption to avoid paying a high federal tax (compositions must be approved by the U.S. Treasury Department). Isopropyl alcohol (isopropanol) in flexo inks is primarily used in polyamide and water-based formulations. Normal propyl alcohol is used to slow down the evaporation rate of inks containing more volatile alcohols.

Glycol Ethers

Glycol ethers are alcohols with ether linkages that make them stronger solvents than either the corresponding alcohol or ether alone. They are excellent solvents for most

Heatset oils

	415 Oil	440 Oil	470 Oil	500 Oil	535 Oil	Magie-Sol 44	Magie-Sol 47	Magie-Sol 52
Specific gravity	.7844	.8026	.8044	.8170	.8373	.7936	.8003	.8146
Pounds per gallon	6.53	6.68	6.70	6.80	6.97	6.61	6.66	6.78
Flash point (°F)	160	190	210	250	245	200	215	266
Flash point (°C)	71	88	99	121	118	93	102	130
Color	OW*	OW*	OW*	OW*	Straw	WW*	WW*	WW*
Odor	Dist*	Mild	Mild	Mild	Dist*	OL*	OL*	OL*
K.B. number	30.1	27.6	27.0	24.2	26.4	26.7	25.6	22.8
Aniline point	165	175	175	186	174	181	185	197
Composition								
Saturate %	86.4	87.1	84.0	84.2	83.0	100	100	100
Aromatics %	10.3	10.0	11.7	11.7	17.0	0	0	0
Olefins %	3.3	2.9	4.3	4.1	0.0	0	0	0
Autoignition (°F)	525	530	530	625	685	530	530	700
Autoignition (°C)	274	277	277	329	363	277	277	371
Distallation (°F) Data								
Initial	402	438	464	527	511	440	464	530
50%	432	464	489	549	552	457	486	551
End point	490	496	516	578	593	490	525	590
Distillation (°C) Data								
Initial	206	225	240	275	266	227	240	277
50%	222	240	254	287	289	236	252	288
End point	254	275	269	303	312	254	274	310

———

*Abbreviations:
OW—off-white
WW—water white,
Dist—distillate
OL—odorless

Source: Magie Bros. Oil Company, Division Pennzoil Company

synthetic resins. The glycol ethers used in printing inks are soluble in both water and oils.

Ethylene glycol monomethyl ether (methyl Cellosolve), used for flexographic or gravure inks, is a solvent with a slow evaporation rate. Ethylene glycol monoethyl ether (Cellosolve) is chemically similar but evaporates slower. Ethylene glycol monobutyl ether (butyl Cellosolve), which is used to formulate screen inks, evaporates even slower. These glycol ethers are good solvents for nitrocellulose and many other resins, such as shellac, vinyl acetate, and maleic and rosin derivatives. Glycol ethers have special toxicity problems and require special handling.

Ketones

The ketones are strong solvents. Acetone is the simplest, most volatile ketone. In decreasing order of evaporation rate and increasing order of molecular weight (complexity of structure) are methyl ethyl ketone (MEK), methyl isobutyl ketone (MIBK), and cyclohexanone.

Acetone is used in nitrocellulose lacquers and gravure inks. It is miscible with water, alcohols, ethers, oils, hydrocarbons, halogenated hydrocarbons, fatty acid esters, and most organic solvents. Methyl ethyl ketone and methyl isobutyl ketone are both used in gravure and screen inks. Cyclohexanone is used in the formulation of some screen printing inks. Isophorone, less volatile than cyclohexanone, is also used in screen inks.

Ketones are rarely used in flexographic printing inks because they swell rubber and plastic plates. Occasionally, acetone or methyl ethyl ketone (MEK) is used when a powerful, rapidly evaporating solvent is required.

Esters

Esters are usually made by reacting an acid and an alcohol. When acetic acid is used, the esters are called acetates. Ethyl acetate is used extensively in flexographic and gravure inks as a rapidly drying solvent. Isopropyl acetate evaporates more slowly than ethyl acetate but more quickly than n-propyl acetate or n-butyl acetate. Esters are generally good solvents for cellulose nitrate.

Nitroparaffins

The nitroparaffin 2-nitropropane is an excellent solvent for reducing the viscosity of most fluid inks. It promotes good adhesion to plastic films; however, its flammability and particularly its toxicity have limited its use.

Additives

The suspension of pigment in varnish and/or solvent does not usually provide a satisfactory ink. Many other materials must be added in order to provide good performance. These additives include plasticizers, wetting agents, antisetoff compounds, waxes, shortening compounds, reducers, stiffening agents, antiskinning agents, and antipinhole compounds.

Plasticizers

Plasticizers make the ink softer, more flexible, and more adherent to the substrate. Resins are often stiff and brittle. Thus, they adhere poorly to flexible substrates. A plasticizer, by improving the plasticity of the film, improves adhesion. Selection of the plasticizer depends on the printed product and the resin in the ink. Since the boiling points of plasticizers are generally high, they do not evaporate in the dryer, and they become a permanent part of the ink film.

Plasticizers include various phosphates, sulfonamides, chlorinated materials, citrates, adipates, polyglycol derivatives, resinates, and phthalates. They are commonly used in flexographic and screen printing inks. Trioctyl phosphate is used in gravure inks.

Waxes

Most printing inks contain waxes to improve rub and scuff resistance. Waxes are used in the form of a compound, which is a fine dispersion of several waxes in an appropriate oil vehicle. They are also used in the form of a micronized, dry powder.

Common classifications of waxes are animal (beeswax, lanolin), vegetable (carnauba, candellila), mineral (paraffin, microcrystalline), and synthetic (polyethylene, polyethylene glycols, Teflon). Incorporated into the ink, wax particles are held to the printed surface by surface tension. If a wax is too soluble in the ink, it will come out of solution when the ink cures, form a waxy film on the surface, and prevent the wet ink from trapping on the dry ink film. This problem is commonly referred to as **crystallization.**

Carnauba wax is especially hard and enhances slip in inks and overprint varnishes. The hard, waxy surface of dried ink films containing carnauba resists marring and scratching. It will also resist dry trapping if the wax is dissolved instead of dispersed.

Petroleum waxes (mineral types) are extracted from crude petroleum. They range from soft to hard, depending on their chemical structure. High-melting waxes are usually hard; low-melting waxes are often soft. They are used primarily to reduce the tack of letterpress and lithographic inks and as slip agents. A petroleum wax of very fine crystalline particle size, microcrystalline wax, is particularly effective as a slip agent. Petroleum jelly, a mixture of liquid and solid hydrocarbons with a consistency similar to grease, retards the drying speed of letterpress and lithographic inks on the press rollers. It is also used to reduce ink tack, improve ink setting or penetration into the substrate, and reduce linting in news inks.

Fine dispersions of polyethylene waxes or PTFE (Teflon) are added to inks to enhance slip and rub resistance. Polyethylene wax is tougher than other waxes, such as microcrystalline wax, but it may lack high surface slip. PTFE produces maximum surface slip and good heat resistance in heatset inks. Excessive use of wax should be avoided to prevent problems associated with ink softening and improper ink viscosity. Increased drying time, trapping problems, and especially decreased gloss are caused by too much wax in the ink. Optimum properties are normally achieved by incorporating about 3% wax, based on total formula weight.

Wetting Agents

Wetting agents promote the dispersion of pigments in ink varnishes. There are many varieties of wetting agents, and the selection of an appropriate one should be left up to the ink manufacturer. Wetting agents for lithographic inks must be carefully selected; otherwise they can cause excessive emulsification of dampening solution into the ink and associated problems. These problems may be resolved using water-pick-up control agents. Wetting agents may also reduce the surface tension of aqueous inks in order to improve the wetting of plastic and foil substrates.

Antisetoff Compounds

Setoff is often referred to as offsetting, but GATF prefers to reserve the word "offset" for a method of printing. Various compounds prevent setoff either by protecting the ink surface or by shortening the ink (decreasing its gelling time). Compounds containing wax or grease shorten the ink, thereby speeding up its setting. Antisetoff compounds that contain magnesia also prevent setoff primarily by shortening the ink; however, addition of such materials often promotes blanket piling.

Starch reduces setoff by slightly roughening the surface of the ink film. The starch is blown onto the printed ink; it should never be mixed with the ink. Starch reduces contact between adjacent sheets and allows more time for the ink to set.

Solvents that increase the rate at which a paper absorbs an ink vehicle may also reduce setoff. However, they tend to reduce the gloss of the dried inks and may also cause chalking on coated stocks. Wax, starch, and magnesia compounds are less objectionable. The wax compounds, if used in excess, can prevent good trapping of succeeding colors if the inks dry between printings.

Shortening Compounds

Shortening agents reduce ink flying, or misting. The addition of a wax compound shortens an ink. The printer should not add such materials except on the advice of the ink manufacturer because they can interfere with proper ink flow on press and aggravate ink piling.

Reducers

Heatset oils or other petroleum solvents are occasionally added to soften and reduce the tack of an ink. A light varnish, such as #0000 litho varnish, boiled linseed oil, or a light linseed isophthalic alkyd, will also reduce the tack of an ink. It is much easier to reduce the tack than it is to increase it. Presently, this is possible using tack increasers.

Stiffening Agents

For sheetfed inks, body gum—#8, #9, or #10 linseed varnish—is used to stiffen an ink that is too liquid and fails to print cleanly and sharply. It "pulls the ink together" when the ink tends to cause scumming or tinting, and it can help prevent chalking on coated stocks. Heavy-bodied gloss varnish (binding varnish) or gel varnish can also be used for this purpose.

Antiskinning Agents

Antiskinning agents are antioxidants that counteract the drying of sheetfed offset inks so that they will not skin over in the can. If they are sufficiently volatile, they will not greatly retard the drying of the rest of the ink.

Printers sometimes spray antioxidants onto the plate and blanket of a sheetfed press to prevent drying or skinning during lunch time. The ink so treated will not dry properly, but a few sheets run through the press will take most of the treated ink away.

Defoamers for Aqueous Inks   Defoamers for aqueous inks are surface-active blends of hydrocarbon liquids, surfactants, metal soaps, hydrophobic silica, and other ingredients, with or without silicone modification. They may be used during the preparation of aqueous inks to prevent foam buildup and may also be added during application for the same purpose. Interactions between waxes and defoamers are sometimes encountered, requiring changes in either the wax or the defoamer.

7 Driers

Although driers are used only in sheetfed printing, they are so widely misunderstood and cause so much trouble that they deserve special attention.

In order to dry in a reasonable amount of time, inks that contain vehicles or varnishes prepared from drying oils such as linseed or tung oil need a catalyst or **drier.** Drying of such inks is a complicated chemical process that occurs very slowly, if at all, in the absence of a drier. Inks do not dry in the absence of air, which explains why inks do not dry in a closed can.

Double bonds (chemically reactive sites in the varnish molecules) readily add oxygen to form compounds called peroxides or hydroperoxides. Peroxides are not especially reactive at room temperature, and when they do decompose, they often return to the original double bond, releasing oxygen. To get these hydroperoxides "moving" (and in the right direction), the ink manufacturer adds metal salts—cobalt, manganese, zirconium, and others—that decompose the hydroperoxide and form a free radical.

The free radical that is formed is called an allylic free radical, and, as free radicals go, it is not especially reactive, although it is a great deal more reactive than the hydroperoxide. This explains why sheetfed inks take several hours to dry, while ultraviolet and electron beam inks, which generate acrylic free radicals, dry in a few seconds.

The allylic free radical can react with many components in the ink. However, when it finds another double bond, it forms a new chemical bond that links it to the new molecule; it also forms a new free radical that can undergo another reaction. This type of reaction in which one free radical reacts to form a new free radical is called a **chain reaction.**

Chain reaction

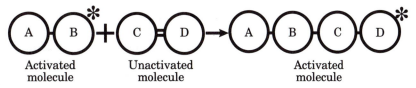

Because this reaction is so complicated, and because it is affected by so many variables, GATF urges printers to avoid altering inks in the pressroom. It is much more convenient to get press-ready canned ink, and, with good planning, it is usually possible to order it. When changes are required, they should be done with the ink manufacturer's advice, using approved materials.

Metals

Salts of many metals will catalyze the oxidative/polymerization reaction by which sheetfed inks dry. Among these are cobalt, lead, manganese, iron, zirconium, and cerium. Mixed driers are more effective than single driers. For years, the conventional drier was a three-way drier composed of cobalt, lead, and manganese. Since environmental regulations severely restrict the use of lead, two-way driers (cobalt and manganese) have become increasingly common and three-way driers using zirconium instead of lead are occasionally used.

Cobalt is a very active drier. It is referred to as a **top drier** since it gives a very hard surface to the ink. The high activity causes problems such as ink crystallization and gloss ghosting if the drier is not used carefully.

Manganese is less active than cobalt. It is referred to as a **through drier** because it dries the ink film throughout and does not form a hard surface. Its dark color may affect the color of white or yellow inks and tints.

Lead salts are even less active than cobalt, but they are colorless. Because of the toxicity of lead salts and because the ink manufacturer cannot always be sure just where the printed ink may end up, these are rarely used.

Cerium, zirconium, and lithium salts are of medium efficiency and, hence, find very limited use in printing inks. Their pale color makes them useful in whites and tints, and they sometimes replace lead in three-way driers.

The salt chosen not only affects the rate of drying and the ease of mixing, it can also create drying problems and print odor, which is a primary consideration in food packaging.

Inorganic peroxides act as driers by promoting the formation of hydroperoxides so that the metal salts can find more sites at which to react. For the ink manufacturer, peroxides are hazardous, flammable products that require the utmost caution in handling until they are finally diluted in the ink. They also tend to produce skinning problems.

Calcium perborate (Graphodrier from Sweden) releases oxygen in the ink fountain to aid drying. The perborates dissipate when mixed with ink in the can. They work particularly well on harder, less porous substrates.

Liquid Driers When the metal salts are suspended in a liquid such as a petroleum solvent, they are referred to as liquid driers. The salts that are soluble in petroleum solvents or oils are the octoates, naphthenates, linoleates (from linseed oil), and resinates (salts of rosin acids).

Paste Driers Most printing ink driers consist of soluble driers containing resins and plasticizers to achieve the desired body or viscosity. Printing ink driers modified with lead linoleate or resinate were once commonly used paste driers. The paste driers in use today are soluble driers cooked with resins to yield a controlled high-viscosity drier that can be added to a printing ink without destroying its body.

Inhibitors Natural drying oils usually contain some drying inhibitors—reducing compounds—and therefore addition of small amounts of drier are ineffective in promoting drying. Once the inhibitor has been overcome, very little additional drier is needed to have optimum effect. The ink manufacturer will have incorporated sufficient drier to overcome the inhibitor. Should the printer be required to add more drier, the presence of inhibitor can be ignored.

Carbon blacks, being very fine pigments, inhibit drying by adsorbing drier from the ink. The ink manufacturer, therefore, adds more drier to black ink than to other inks.

Effect of drier content on drying time of ink

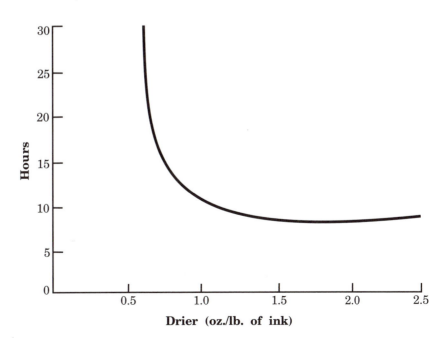

Accelerators   Cobalt acetate is sometimes added to the dampening solution of the offset press to accelerate ink drying. This has much the same effect as adding more drier to the ink. It has the disadvantage of being hard to control.

Mixing   It is a popular misconception that the rollers on the offset lithographic (or letterpress) press will properly mix any material added to the ink in the fountain (duct). The inking system is designed to deliver a uniform ink film to the plate, not to mix ink additives.

If driers or other materials must be added to the ink, the ink should be removed from the fountain and weighed. The proper amount of additive should be weighed, added to the ink, and thoroughly mixed on the slab before returning the ink to the fountain.

The ink manufacturer has convenient mixing and weighing equipment and can make proper additions less expensively than the press crew can correct improper ones.

Effect of Temperature and Acidity   Increasing the temperature increases the rate of most chemical reactions, and this is also true of ink drying. However, increasing the temperature also increases the fluidity of the ink, and this effect is much greater than the effect on drying. Uncontrolled increases in temperature can cause poor dot formation, fill-in of shadow dots, blocking,

Effect of temperature on ink drying time

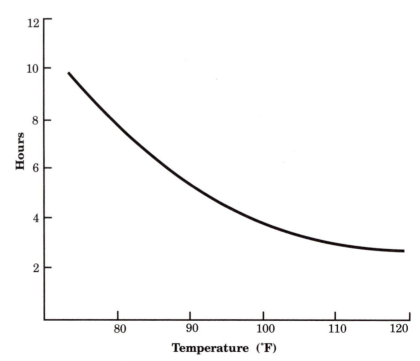

Temperature (°F)

and other printing and drying problems. (Spontaneous ignition of nitrocellulose-coated book covers has been reported.) Use of heat to accelerate drying must be controlled, as must everything else in the process.

While the graph shows ink drying at pH 3 and 65% relative humidity (R.H.), the printer should remember that within the pile of printed paper, the R.H. approaches 100%.

Effect of relative humidity on the drying times of three inks

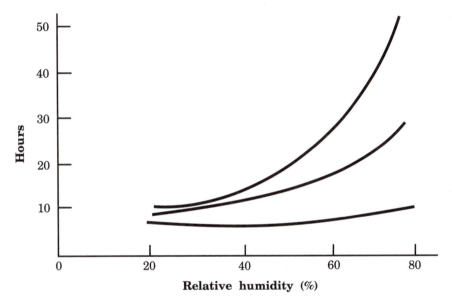

The greatest cause of drying failure in sheetfed printing is the combination of too much water and too much acid. The accompanying illustration presents laboratory results in graphic form.

Effect of relative humidity (R.H.) and changing pH on drying time

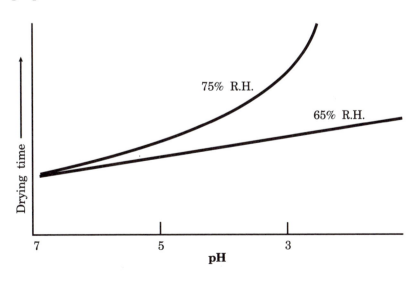

The data suggest that water and acid are combining to hydrolyze the drier salt. In any case, they strongly support GATF's recommendation that dampening solution pH be kept between 4.5 and 5.5. Several commercial etches are available that, when used according to directions, give a pH significantly lower than 4.5. Printers who use them often report more than the usual number of drying problems.

Printers often believe that litho plates run cleaner when the pH is low. This certainly is true of stone plates, zinc plates, and bimetal plates. There is no evidence that a pH lower than 4.5 helps with aluminum or paper plates. In any case, some proprietary dampening solutions give excellent results without using a pH lower than 4.5.

8 How Printing Inks Set and Dry

When printers refer to setting or drying, they are defining the increase in the viscosity or the body of the ink that occurs after the ink is exposed to air. It is very difficult to determine how much of the increase in viscosity is due to setting and how much is due to drying. Setting, which is caused by the solvent evaporating from the ink or being absorbed by the paper, occurs very quickly after the ink is printed. Drying, which is a chemical process, occurs more slowly.

Setting

The role of thixotropy. Paste inks, which are used in lithographic and letterpress printing, become less viscous when they are stirred or worked. This property is known as **thixotropy** and explains why paste inks have a higher viscosity when they are first removed from the can than they do after they have been worked with a knife on a slab or in the inking system of the press. The bonds within paste inks are relatively weak; they are broken simply by stirring. On the press, the ink flows readily, creating a uniform layer of ink on the rollers, plate, and blanket. When stirring, mixing, or working is stopped, the weak bonds reform and the ink becomes more viscous again. Thixotropy thus promotes the setting of ink. After the ink is printed on paper or another substrate, the viscosity of the ink increases.

Gellation. One way of increasing an ink's viscosity so that it will set is to form a gel. In the coatings industry, it is possible to disperse a dry resin into a plasticizer, coat a sheet with it, and cause the resin to form a gel by heating it. These suspensions, called plastisols, are used in printing textiles by the screen process.

A similar procedure is used in making quickset ink. A high-viscosity varnish can be diluted with a solvent so that it will flow easily. This is called a **quickset varnish.** When an ink containing a quickset varnish is printed onto a paper, the paper absorbs the solvent, and the viscosity rises rapidly, setting the ink. A quickset ink does not dry on nonporous material, such as foil or plastic, because the oil is not absorbed and the ink viscosity cannot increase.

Drying

Drying is the most important property of ink. If the ink does not dry, it is useless. There are several different mechanisms by which inks dry. In practice, many of these occur simultaneously in the drying of most inks.

Absorption. Absorption occurs whenever ink is printed on paper. The solvents in gravure, flexo, heatset, and even

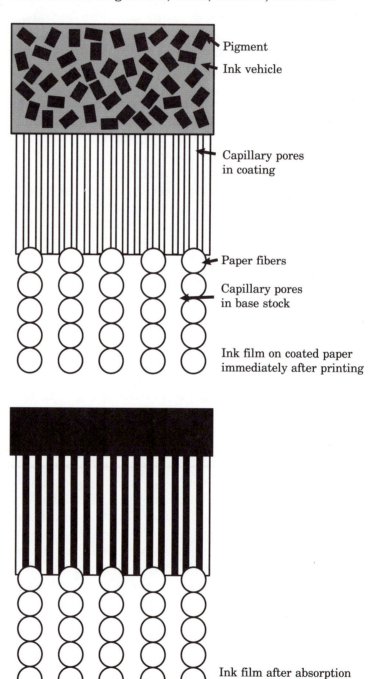

An ink film setting on coated paper

Pigment

Ink vehicle

Capillary pores in coating

Paper fibers

Capillary pores in base stock

Ink film on coated paper immediately after printing

Ink film after absorption of vehicle in the coating

sheetfed inks penetrate paper to some extent, and the viscosity of the ink remaining on the surface of the paper increases somewhat. News inks, however, dry solely by absorption. When the newspaper is printed, the oil very quickly drains out of the ink, leaving the pigment more or less attached to the newsprint.

Absorption also helps in the setting and drying of other inks. Closely related to news inks are the nonheatset web offset inks that dry by absorption.

An ink film setting on uncoated paper

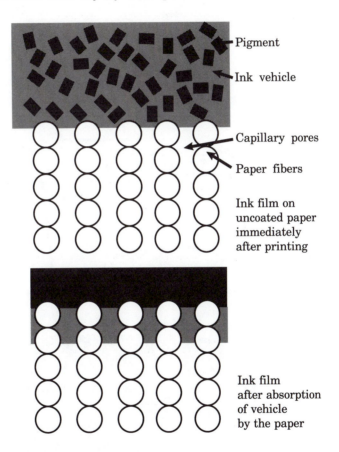

Pigment

Ink vehicle

Capillary pores

Paper fibers

Ink film on uncoated paper immediately after printing

Ink film after absorption of vehicle by the paper

Evaporation. Except for newspaper, forms, and nonheatset web offset inks, web inks dry by evaporation. Whether it is water or alcohol evaporating from a flexo ink, an ester or hydrocarbon from a gravure ink, a heatset solvent from a heatset web offset ink, or water or solvent from a screen ink, drying of the ink is accomplished by evaporating the solvent. The resin left behind then bonds the pigment to the paper or film so that it will not rub off.

Evaporation occurs very rapidly, permitting web heatset to run at speeds of 2,000 ft./min., or higher for flexography, and as high as 2,500 ft./min. for gravure.

Control of drying is of utmost importance if the print is to be delivered without "cooking" (overheating) the paper or film. Modern high-velocity hot-air (HVHA) dryers operate more efficiently than older ones because of engineering design advances. They are far more efficient than the old flame-impingement dryers first used for web offset.

Putting a controlled ripple in the web
Courtesy TEC Systems, Inc.

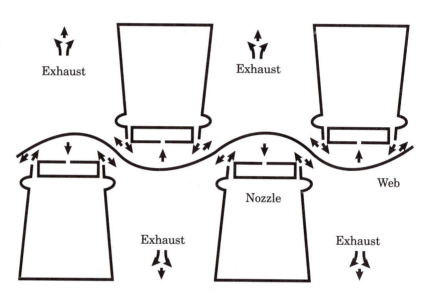

High-velocity hot-air dryer

Use of an optical pyrometer helps the operator to control drying. The optical pyrometer senses the temperature of the moving web. Evaporation is better controlled by measuring the temperature of the web (from which the ink evaporates) than by using a thermocouple to measure the temperature of the hot air used for drying.

THERMALERT®
optical pyrometer
*Courtesy Raytek
Inc.*

Oxidative polymerization. Many details of the reaction are still not fully understood. Despite the complexity of this chemical reaction, it rarely fails to occur, and the ink almost always dries.

The principle of polymerization is illustrated in the diagram on page 92.

Varnishes that are derived from drying oils, such as linseed oil, chinawood oil, and/or soya bean oil, react with oxygen in the air. In the presence of a cobalt or manganese salt (called a drier, initiator, or catalyst), the reaction forms a molecular chain that continues to grow. This process is called crosslinking. As the chain gets longer and longer, it flows less and less readily until, after 2–4 hr.,

Polymerization

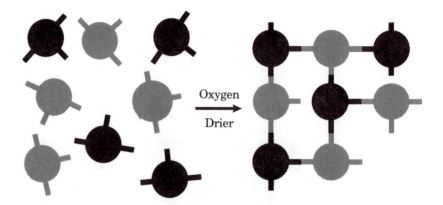

enough chains have been formed to stop the ink from flowing. It will not smear when it is rubbed; it is dry.

It is still not fully understood why cobalt salts act as top driers, which promote rapid drying of the top of the ink film, while manganese salts act as "through" driers, which promote drying of the ink film throughout. Good lithographic inks must contain at least two types of driers. Lead salts were favored for many years but are no longer used because of their toxicity.

The drier is a critically important ink constituent. In its absence, the ink dries very slowly, if at all. In lithography, the pH of the dampening solution must be carefully regulated. Among other things, excess acid in the dampening solution (pH too low) can inactivate the drier.

Because of the nature of oxidative polymerization, the printer must take special precautions. The most important ones are:
- Be sure to select an ink suited to the substrate.
- Do not alter the ink without consulting the ink manufacturer.
- Use a minimum amount of dampening solution and keep the pH of it between 4.5 and 5.5. Low pH inhibits the reaction by which these inks dry.

Vehicles based on drying oils are sometimes added to web inks to provide extra hardness after setting. For example, many web offset inks can be smeared if handled with greasy fingers. However, if a drying varnish is used, the dried ink becomes much harder after a matter of hours.

Radiation polymerization. Unlike oxidative polymerization, which requires 2-4 hr., radiation polymerization is complete in a fraction of a second. Under the influence of ultraviolet light or electron beam radiation, inks containing special acrylate monomers react very quickly. In a matter of seconds, prints are dry and ready to be cut or folded. Spray powder is not required to prevent setoff, and with folding cartons, the resulting varnished print has a smoothness that cannot be matched with conventional sheetfed inks.

These inks are more costly than conventional sheetfed or web inks, and their application has been largely limited to metal can decorating, where the quick drying speeds production; the printing of folding cartons that command a premium price, such as cosmetic boxes and gift boxes for liquors; album jackets; and screen printing, where the speed of the reaction is useful in printing plastic containers.

Catalytic polymerization. When the substrate will withstand the heat, as with metal cans or glass, a slower-reacting polymerization can be used to cure the ink. For example, melamine formaldehyde polymers result in a hard, resistant film of ink or varnish on beverage cans and bottles. To achieve this hard film, a heat-reactive melamine formaldehyde prepolymer is mixed with a catalyst. On baking, the catalyst initiates the polymerization. Several other catalyzed polymerizations are also used for baking varnishes and inks.

Infrared radiation. Infrared radiation, as a source of heat, promotes absorption, evaporation, and oxidative polymerization. When inks are heated, they usually dry faster. Although heat reduces the viscosity of the ink (slows drying), it also accelerates the absorption and evaporation of solvent from quickset inks, and accelerates the oxidative polymerization reaction, so that, when the infrared unit is kept under control, setting and drying occur faster.

Because infrared radiation accelerates drying, prints are usually ready to be backed up in a few minutes and can be ready for cutting and trimming in less than an hour.

The description of "tuned IR dryers" is often misleading because some IR dryers emit most of their radiation in the near infrared while others emit mostly in the far infrared. The near infrared is the most energetic (it dries the fastest), but it is also very selective of color; that is, the black areas of the image are dried before the yellow begins to dry. The far infrared dries all colors evenly, but very slowly. The middle range seems to be the wavelength selected by most printers who use infrared to speed drying and reduce spray powder.

Infrared radiation can also be used with high-velocity air for web printing. By applying infrared radiation to the print, then using high-velocity air to remove the solvent, the printer can dry the web at a cost only slightly higher than that of gas drying.

Inks that dry by several mechanisms. Water-based flexo news inks dry by a complex combination of mechanisms: penetration of water in the newsprint; evaporation of water, ammonia, or amine; and decomposition of the resin/ammonia salt (ink binder). Because a continuous film is formed, this method of drying is conducive to good rub resistance. This method also maintains the opacity of the newsprint and does not contribute to show-through.

Show-through with letterpress and web offset oil inks is caused by the mineral and ink oils that penetrate the interspace among the fibers and replace part of the air. Since the refractive indices of the oil and of the paper fibers are nearly identical, the paper becomes transparent and causes show-through. The ability to print lightweight newsprint is a major advantage of water-based flexo inks.

A review of the mechanisms discussed in this chapter reveals that an ink may set and dry by several different ways. Consider a heatset web offset ink that is applied to a fast-moving web of paper. When the ink first touches the sheet, some of the solvent is absorbed. Absorption of the solvent initiates gellation, which greatly speeds the setting of the ink. The print now goes into the dryer, where solvent is evaporated, and then it passes over the chill roll, where the softened resin is cooled and set. If the ink contains a drying oil, final hard drying will occur by oxidative polymerization.

Deinking

This chapter has been devoted to the process of converting a liquid film into a solid. Sometimes it is desirable to go the other way. The value of wastepaper increases if the dried ink can be removed. Between a quarter and a third of the paper and board manufactured in the United States is recycled, a greater percentage in Europe and almost two-thirds in Japan.

Some wastepaper is merely dispersed in water and reformed into sheets. Chipboard, the gray material on the back of writing tablets, is treated this way. If the paper or board was unprinted in the first place—for example, trimmings from envelopes or paper dinner plates—it can be used to make white sheets without deinking.

Most recycled waste, however, must go through a deinking treatment, usually with caustic, to soften the ink and remove it from the paper fibers.

Paper and board printed with ultraviolet-curing inks can also be deinked, but it is more difficult and requires a costly process that reduces the value of the waste.

Different processes are used for deinking paper, and some are more suitable than others. Deinking processes are available for most kinds of ink, but mixed wastepaper that contains different kinds of ink can pose problems.

9 Ink Manufacture

To provide full service, ink manufacturers must know the properties of pigments, vehicles, solvents, driers, and additives, as well as how to select and combine them to produce an ink with the required runnability and end-use characteristics. They must also know surface chemistry, rheology, the mechanics of printing ink transfer, and the latest technology in pigments and vehicles. In light of this, ink manufacturers seem to bear full responsibility for formulating the correct ink for a particular printing process and substrate. However, they do not; printers also have a very important role in ink manufacture. Printers must provide ink manufacturers with the necessary information about a job in order to expect formulation of a satisfactory ink.

In order to formulate an ink that will perform properly for a special job, the ink manufacturer must first get all of the job specifications: color, printing process, press, paper or other substrate to be used, and any other requirements. After the formulation is worked out and tested, the ink manufacturer prepares the amount of ink that has been ordered and then tests it again to make sure that it still meets the requirements for the job before presenting it to the printer.

Stock, or off-the-shelf, inks are prepared according to standard formulas. Stock inks will often meet the printer's requirements for a special job. It is the ink manufacturer's responsibility to know whether a stock ink will work well or whether a special ink is needed.

Consider, for example, the formulation for a heatset web offset publication ink presented in the accompanying illustration. The ink manufacturer blends two flushed pigments with heatset varnish, heatset gloss varnish, and slip compound. After thorough mixing, the product is milled on a three-roll mill. Then it is tested in the laboratory, and an appropriate amount of heatset solvent is added to adjust the tack. It is tested again and then packaged. In many cases, the ingredients are dispersed enough for the ink manufacturer to omit three-roll milling and, instead, do a simple filtration of the finished ink.

Referring again to the illustration, note that the flushed pigments are formulated from pigment, a modified rosin binder, and a solvent. Two pigments are chosen to give the proper compromise between cost and color; for example, a warm red may be blended with a magenta. The amount of

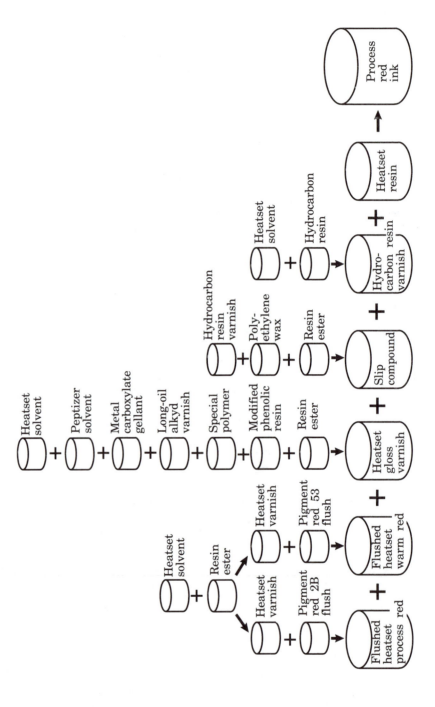

Components of one heatset ink

pigment in each flush and the amount of flush used in the ink must be carefully controlled if the ink is to have the right hue, gloss, and color strength.

The heatset gloss varnish is itself a complex formulation. The ink manufacturer may formulate it or buy it from a varnish maker. In addition to heatset solvent and rosin ester, the heatset gloss varnish contains a long-oil alkyd (a drying oil) so that the ink will dry hard after it has been printed. (Less expensive inks that do not contain drying oil will soften and smear if they are brushed with oily or greasy fingers.) The heatset gloss varnish also contains a peptizer solvent to assist in dispersing the pigment, a metal carboxylate gellant to provide quickset properties, and a modified phenolic resin to provide gloss.

The slip compound improves the rub resistance of the ink. It consists of a wax dispersed in a rosin ester varnish plus a hydrocarbon resin varnish. Teflon and other plastic powders may also be added to improve slip.

The first five ingredients (two pigments, gloss varnish, slip compound, and hydrocarbon resin) are thoroughly mixed together in order to make the finished ink. This mixture is sent to the lab for evaluation and then let down with heatset solvent to adjust the tack. The ink may be given a touch-up grind before packaging or it may be filtered.

Inks can be as complex as this one or as simple as a news ink, which is a blend of two mineral oils (to achieve the right body or viscosity), a black pigment, and a small amount of toner to give it a blue-black instead of a brown-black shade.

Liquid inks also vary in complexity. The simplest liquid inks are prepared by dispersing chips (pigment dispersed in a solid resin) into the solvent and milling.

Paste and Liquid Inks

Inks can be classified as either paste or liquid (fluid). Paste inks are used for litho, screen, and letterpress printing. Liquid inks are used for flexo and gravure printing. News inks, although they have a lower body than commercial offset and letterpress inks, are commonly considered to be paste inks. The ingredients and manufacturing procedures for these two types are significantly different.

Liquid Inks

Processing of liquid inks. Flexo and gravure inks are easily processed because of their low viscosity. However, they are usually processed in closed containers. The solvents must be highly volatile to dry properly at the low temperatures used for drying on the gravure or flexo presses.

Historically, the ingredients, which consisted of pigment(s), solvent(s), and resin, were mixed in a ball mill. The mixture was then combined with ceramic balls in a large cylindrical vessel or crock. The vessel was tightly closed, turned on its side, and rolled on rollers to mill the ink.

These inks are now manufactured in continuous mills in which the batch flows through a stirring medium consisting of small steel or glass balls or, in rare cases, sand. Because low viscosity allows pigments to settle out, liquid inks are usually formulated at concentrations higher than those needed on the press. The printer dilutes the ink

Bead mill
Courtesy Buhler-Miag, Inc.

with a suitable solvent at press side. This practice also reduces shipping and storage costs.

Gravure inks. In the manufacture of gravure inks, certain resins are dissolved in solvents and milled with the pigment. The product is filtered, and the refined ink is ready for the printer. Since volatile solvents and flammable resins, such as cellulose nitrate, are involved, provisions must be made against possible fire hazard.

Flexographic inks. Procedures for manufacturing flexographic inks are similar to those for gravure inks. Flexographic inks, like gravure inks, can be produced from chips, which are dispersions of pigments in a dry resin. Since most of the work required to disperse the pigment was done in preparing the chips, only a high-speed mixer is required to make ink from chips.

Technological advances in aqueous liquid inks have opened new markets for water-based flexo inks. Water-based flexo inks contain acrylic resins which give them good pigment wetting characteristics and excellent surface holdout and opacity. Newspapers, previously printed by letterpress, are now being printed by offset lithography and flexo. Flexo is expected to capture even more of this market because, with water-based flexo inks, good surface holdout and opacity can be achieved with a lighter-basis-weight newsprint than that used with offset.

Paste Inks

Paste inks may be made by grinding the dry pigment into the varnish and dispersing the other additives required to modify flow, tack, and film properties. Generally, the thin litho varnishes (up to #000) wet pigments poorly and tend to produce short, nontacky inks. These varnishes also tend to penetrate paper rapidly, leaving the ink film with insufficient binder. The medium varnishes (#000 to #2) have good wetting properties for pigments and produce inks with good length and tack. They are the most commonly used varnishes. The heavy-body varnishes (#3 to #8) are used to increase tack, pigment binding, and water resistance.

Varnishes #9 and #10 are often called binding varnishes or body gums. In general, pigments that are easily wet can be ground in thinner varnishes than those that are hard to wet.

Most modern paste ink varnishes have resins (either rosin derivatives or hydrocarbon types) added to the drying oils to improve drying, gloss, and lithographic properties.

The selection of pigment, vehicle, and grinding equipment depends on the particular printing process. Specific formulations of inks for the various printing processes are given in chapters 12–16. Some general procedures in making inks are as follows.

Lithographic and letterpress inks. Paste inks are used for offset and letterpress printing. The differences between lithographic and letterpress inks are primarily in their formulations (pigment concentration and choice of pigment and varnish) and not in the method of manufacturing them.

Newspaper inks. Black newspaper inks are basically comprised of lamp black pigment and a small amount of blue toner dispersed in mineral oil. The pigment content is low, normally about 10%.

Color in newspapers is becoming increasingly popular. Colored news inks used for printing in the main body of the paper are known as "ROP" (run of press) inks. Most ROP inks are produced by blending mineral oil with pigment flushes, followed by filtering the finished ink.

Low-rub news inks are becoming more popular because they contain drying oils or resins to adhere the pigments to the stock. Conventional news inks do not contain a binder.

Screen printing inks. Screen printing inks usually dry by a combination of solvent evaporation and oxidation. They have a low tack and a thick, nonfluid body so that they can be forced by a squeegee through the stencil screen. Screen inks are similar to paste inks that have been greatly reduced with mineral spirits.

UV inks that dry upon exposure to ultraviolet light are widely used for screen printing.

Mixing and Milling

In making an ink, the ingredients are first mixed thoroughly and then milled to complete the dispersion of the pigment and take the air out. However, with the development of more easily dispersed pigments and pigment dispersions or flushes, the boundary dividing mixing and milling is not so clear.

Commonly used mills and mixers include ball or pebble mills, sand mills, roll mills, stone (carborundum) mills, the Kady mill, the Cowles disperser, and dough and Banbury mills.

Twin-motion mixer
*Courtesy Day
Mixing Co.*

The Banbury and high-speed disperser mixers handle batches of ink, while the sand mill processes the ink continuously. A sand mill is a vertical device with a rotating shaft that has flat impellers. A **premix** is pumped into the bottom of the grinding chamber, which is filled with a special sand or other grinding media. The product is screened at the top of the mill to remove abrasive particles.

In making paste inks, the ingredients are ordinarily dispersed using a high-speed disperser mixer, then milled in a three-roll mill. Liquid inks are too volatile and too low in viscosity to be manufactured with this equipment.

Twin-shaft
disperser
*Courtesy Buhler-
Miag, Inc.*

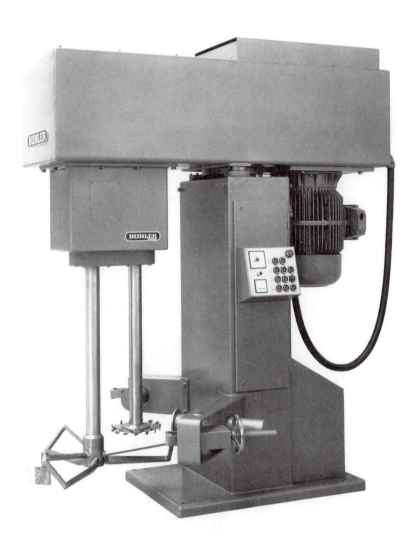

**Three-Roll
Mill**

The three-roll mill generally handles high-viscosity paste inks made from pigments and heavy varnishes. The rolls rotate in opposite directions at differential speeds. The paste ink is fed between the back roll and the center roll and passes through the nips to the front roll, where it is removed with a doctor blade. As the ink passes between the nips, it is sheared or ground and the pigment particles are thoroughly covered with varnish. The nip also withholds or classifies the coarse particles by actually pushing them to each side.

The ink must be carefully milled to completely disperse the pigment and give the ink full color value. Without proper milling, the ink will perform poorly on the press.

Programmable
three-roll mill
*Courtesy Buhler-
Miag, Inc.*

Operation of the three-roll mill is a craft. The adjustment of the pressure between the rolls must be exactly right. If it is too great, throughput is reduced and the mill tends to run too hot. If the roll is run too loose, the ink is not ground properly and does not develop full color value or strength, although throughput is greatly increased. Since throughput of a properly operated three-roll mill is not very great, and since a skilled operator is required, milling is an expensive operation. The throughput can be greatly increased (and costs reduced) by reducing the pressure between the rolls, but the resulting ink will have less than the full color value. Since color strength is usually the most valuable feature of the ink, poorly ground ink is not a bargain.

Grinding action of
a three-roll
dispersion mill
*Courtesy Day
Mixing Co.*

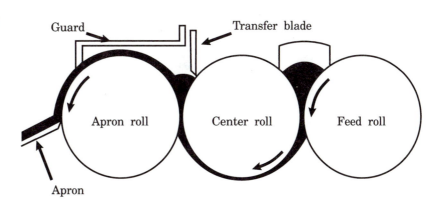

Guard   Transfer blade

Apron roll   Center roll   Feed roll

Apron

10 Testing

GATF recommends that printers routinely check their inks at predetermined intervals, especially when the equipment or other printing products are changed. Printers should also invest in some simple ink testing to determine whether the inks that they purchase will perform as expected. Unique or unusual tests may be warranted in special circumstances. For these, printers normally rely on the ink manufacturer for assistance. GATF recommends that the printer visit the ink plant to become acquainted with the ink manufacturer's testing facilities and procedures.

If the printer has been a good customer, the ink manufacturer will usually perform some basic ink tests. However, testing for color strength, tack, or other properties that require special skills and equipment is costly, and the ink manufacturer is justified in referring the customer to a special lab for extensive testing. (GATF and other testing laboratories can perform a wide range of ink tests.)

In developing a testing program, the printer should retain a consultant (GATF or the ink manufacturer can often be of assistance) to help determine which tests are likely to be useful and what equipment is required for them. Much of the initial testing will be carried out by the consultant or commercial laboratory. As the program develops, it will become apparent that some tests can be performed more conveniently or economically in the printing plant than by a testing laboratory or consultant. Only a few very large printers can perform chemical tests of ink as economically as a qualified consultant can.

Testing can vary from careful observation by the press operator at press side to complex chemical and physical analyses. Ink manufacturers use a variety of tests that are largely irrelevant in the printing plant (esterification number, iodine number, aniline point, etc.) to determine the properties of raw materials used in making ink. This chapter is limited to those tests that are likely to be useful to a general commercial printer. The various product resistance tests that are of interest to the package printer are mentioned only briefly in this chapter. (Ink manufacturers normally do not test end-use performance on packaging inks.)

Standardized Ink Tests

Descriptions of tests that are discussed in detail in other sources, such as the American Society for Testing and Materials (ASTM) *Annual Book of Standards* and the ASTM *Paint Testing Manual,* are not included in this chapter. Where an ASTM standard test is available, its designation is listed. Other sources of standard tests are the Technical Association of the Pulp and Paper Industry (TAPPI), the International Standards Organization (ISO), its American affiliate, the American National Standards Institute (ANSI), and, in Great Britain, the Research Association of the Paper and Board, Printing, and Packaging Industries (Pira).

Sampling

If testing is to be meaningful, it must be carried out on a representative sample. Tests that are conducted on nonrepresentative samples are meaningless or misleading. The problem of assuring a representative sample is especially acute in testing press sheets, where considerable variation occurs from sheet to sheet, and in testing raw materials, which may arrive in several different packages, cans, or drums. There is extensive literature on sampling, which is as important as any other aspect of product or raw material testing.

Making Test Prints

Ink testing often involves applying ink to paper or some other substrate. There are several ways to do this. The simplest is to put a dab of ink onto paper and draw it down with a spatula. Producing an acceptable ink drawdown with a spatula requires skill and experience, and even then

An ink drawdown

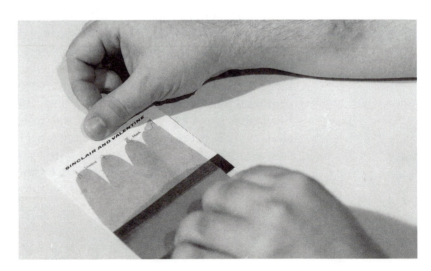

the results are not quantitative. A drawdown knife or spatula is almost useless for applying ink to film, foil, or coated paper. However, relatively simple rollout devices like the Quickpeek color proofing kit for paste inks and the anilox hand proofer for fluid inks permit the operator to apply uniform, reproducible ink films to these and other substrates. At least one of these devices is essential in an ink testing laboratory.

Quickpeek test

Proof presses, such as the Vandercook, Prüfbau, IGT printability tester, and "Little Joe" Offset Proving Press, provide the greatest precision in applying ink to a given substrate for a test print. The Vandercook proof press has been a standard for years.

"Little Joe" Offset Proving Press *Courtesy Graphicart America, Inc.*

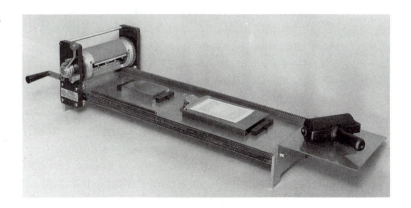

Printability tester
Courtesy Prüfbau
Instrument Co.

The **ink drawdown test** provides information about the color, gloss, and drying of the ink in addition to other printing properties. A drawdown or rollout test is convenient for checking batch-to-batch variations. Undertone, which is simply the color of an ink when drawn out to a very thin film on a white surface, can be checked by either a rollout test or printed proof. Masstone is a heavy ink film that is unaffected by paper; light can pass through it. Both undertone and masstone can be examined with a thick and thin film drawdown. Since a match depends on visual comparison, it is important that the observer be tested for color blindness and other color vision deficiencies. A drawdown test is most easily performed on a swatch pad that can be obtained from any ink manufacturer.

To perform the test, place a small portion of the ink to be tested on the pad next to the ink with which it is to be compared. Use the drawdown knife to draw the samples down to adjoining films, first with heavy pressure for 1 to 1½ in. (25–40 mm), then lightly for another 2 in. (50 mm). The films must be adjacent to each other, but they must not mix. It requires some practice to master this technique.

Judge the masstone of the test sample by comparing the thick films at the top of the drawdown, and the undertone by comparing the thin films at the bottom. View the thin films by reflected light, then hold up the sheet and compare them by transmitted light. Note also any differences in gloss or bronziness. If the drawdown is to be preserved, cover the swatch with transparent cellophane or colorless plastic film. GATF recommends viewing the samples in a color-viewing booth because prints that match under one type of light may not match under another.

Gravure inks should also be examined for middletones. (This technique is also useful for lithographic and other inks.) Make certain that the standard and test inks are of the same viscosity. Place a spot of each ink, about the size of a dime, 1 in. (25 mm) apart on the paper. Make the initial light drawdown with the blade at an angle of about 30° to the paper. Bring the ink samples together and continue to draw down for about another inch. This gives a middletone. At this stage, raise the blade to about a 90° angle and apply greater pressure, drawing down for about 2 in. (50 mm). This gives an undertone. To obtain masstone, continue the drawdown, with the blade again at a 30° angle to the sheet, using minimum pressure so that the blade glides on top of the film. With this procedure, strength and gloss of the undertone, middletone, and masstone, as well as the overall color, can be evaluated.

Optical Properties

Optical properties include color, tinctorial or color strength, opacity or hiding power, and gloss.

Color

One of the prime considerations of an ink is that it be the correct color for the job. Ink color is a combination of masstone and undertone. Both are important in color matching because the color of ink on paper depends on the predominance of reflected or transmitted light. If the film is thick, masstone dominates; if thin, undertone dominates. These two tones are more significant in lithographic ink films than in the thicker gravure or screen ink films. If the ink manufacturer is asked to formulate a new ink, the chemist in the color-matching laboratory will match the shade of the ink by blending colors and comparing a proof with a color sample supplied by the customer. For a perfect color match, the proof must match the standard ink in masstone, undertone, opacity, and gloss. (See earlier section entitled "Making Test Prints" for procedure to determine masstone and undertone.)

Color Measurement

The following ASTM tests are useful in measuring color in the graphic arts.

ASTM color tests relevant to graphic arts

ASTM Number	Brief Title
D-1729	*Visual Evaluation of Color Differences*
D-2244	*Instrumental Evaluation of Color Differences*
D-1535	*Color by the Munsell System*
D-3022	*Color and Strength of Color Pigments*
D-387	*Mass Color and Tinting Strength*
D-2616	*Evaluating Change in Color with a Gray Scale*
D-3134	*Color and Gloss Tolerances of Opaque Materials*
D-1544	*Color of Transparent Liquids (Gardner Scale)*
D-29	*Sampling and Testing Lac Resins*
D-564	*Liquid Driers*
D-156	*Saybolt Color of Petroleum Products*
D-365	*Soluble Nitrocellulose Base Solutions*
D-1925	*Yellowness Index of Plastics*

Color is measured best on a printed sample. A densitometer can be used for a first approximation. By recording the red, green, and blue densities, as measured by a good densitometer, the printer can develop permanent records of the colors of the inks. The *GATF Color Triangle, Color Circle,* and *Color Hexagon* are useful diagrams for recording and analyzing densitometric measurements of colored prints. A description of these devices and their uses is presented in the GATF text *Color and Its Reproduction* by Gary G. Field.

For a complete analysis of color, a spectrophotometer is required. Colorimeters also give a complete description of any color.

Test sample being positioned in spectrophotometer for spectral analysis

**Color
Strength**

Although somewhat troublesome, the test of color strength
(also called tinctorial strength) is probably the most
important ink test for the printer. The value of an ink is
directly related to its color strength. The printer may
request the results of a tinctorial strength test from the
ink manufacturer or get the test from a testing laboratory.
The test is often called a "bleach test." Details of it are
presented in ASTM D-387.

The test requires a white base, usually zinc oxide or
titanium dioxide (for yellows, use a blue tint base—1.5 part
cyan toner to about 1,000 parts white tint base), and an
analytical balance. To ensure compatibility, the tint base
should be selected to match the type of ink to be tested:
sheetfed offset, web offset, gravure, etc.

On the analytical balance, weigh 0.4 g of the ink (0.8 to
1.0 in the case of an opaque yellow) to be tested on a watch
glass. Now add exactly 20.0 g of the tint base ink. Weigh
out exactly the same amount of the standard ink, and add
exactly 20.0 g of tint base. Mix each sample carefully and
thoroughly. Place small portions of each mix side by side
on a test paper with the test sample on the right. Then,
use the drawdown knife to draw them down lightly and
produce adjoining, relatively thick films. Compare the
results immediately (within five seconds). If there is a
difference, judge which tint is stronger. Note also any
difference in shade or cleanness of the tints. If the
drawdown is to be preserved, cover the inked areas with a
piece of transparent film.

For yellow inks, the blue tint base produces greens that
make it easier to judge color strength, as well as slight
differences in hue.

If the percent difference in tinting strength needs to be
determined, make another reduction of the standard ink.
Use more than 20.0 g of the tint base if it is stronger than
the ink being tested, and less than 20.0 g of the tint base if
it is weaker than the ink being tested. To avoid reweighing
the samples, simply add more tint base to the stronger ink.
This procedure gives approximate results and saves time.
When a match has been obtained, calculate the strength of
the test ink, relative to the standard, using the formula:

$$\text{Relative Strength} = \frac{100T}{S}$$

where T is the amount of base ink added to the test ink, and S is the amount added to the standard to produce the match. The relative color of the tints obtained in this test gives a more accurate comparison of undertones of the inks than the ones obtained from a full-strength drawdown. GATF performs this test for its members.

Opacity (Hiding Power)

The test swatches used by ink manufacturers have a black bar across them so that the opacity, or hiding (covering) power, can be judged from the same test used to judge the masstone and undertone. Judge the relative opacities of the standard and sample by the degree to which each hides the underlying black. A more detailed and quantitative test of hiding power is presented in ASTM D-2085.

Gloss

The gloss of two or more inks can be compared visually to judge how well they match. The eye does an excellent job of detecting differences in gloss. Inks should be drawn down, side by side, and allowed to dry.

In comparing different inks for gloss, there are three important considerations:

1. The ink films must be printed on the same stock because the smoothness, absorbency, and gloss of the underlying surface affect the gloss of the printed ink.

2. The ink films must be of equal thickness.

3. The inks must be dry.

If a numerical measurement is required, a gloss meter is used. There are several different meters on the market. Some gloss meters read gloss at one angle only; others can be adjusted to read gloss at several angles. ASTM tests for gloss that are of interest in the graphic arts are listed in the following table:

ASTM tests for gloss

ASTM Number	Brief Title
E-97	*45°, 0° Directional Reflectance of Opaque Specimens*
D-523	*Specular Gloss*
E-167	*Goniophotometry of Reflecting Objects*

Working Properties

The working properties of an ink include drying properties, fineness of grind, wet film thickness, tack, viscosity, flow, length, tendency to fly or mist, tendency to emulsify, pigment bleeding in dampening solutions, scumming tendency, density, specific gravity, and flash point.

Drying

When the printed ink film is dry, the print is ready for the next operation—cutting, folding, binding, or shipping. Several tests indicate what to expect in the workplace fairly well, but no single laboratory test will accurately predict how much time it will take before a printed pile may be cut, folded, or sent to the bindery. Obviously, conditions in the printing plant are among the many factors that affect drying time. For complete information on the rate of ink drying, the ink should be tested both on paper and on an impervious surface such as glass. The time required for ink to dry on paper varies with the stock, and it should be tested on the stock to be printed. The drying time on glass or metal is independent of any stock. It indicates the tendency of an ink to skin in the can and dry on the press rollers. While the ink should dry rapidly on paper, it should stay "open" on the press. Drying time tests are described in the ASTM methods listed in the accompanying table:

ASTM tests for drying

ASTM Number	Brief Title
D-1640	*Drying, Curing, or Film Formation*
D-564	*Testing Liquid Driers*
D-2091	*Print Resistance of Lacquers*
D-1650	*Sampling and Testing Shellac*

Sheetfed inks. To test the drying rate of litho or letterpress inks on paper, the ink film should be about the normal printed thickness. This result is best obtained with a hand proof press or a Quickpeek tester. The volume should be sufficient to produce a film about 0.0004 in. (0.01 mm)—10 micrometers—thick. This quantity will make a film three to five micrometers thick on the proof, depending on the area of the solid.

A simpler method is to make a tap-out of the ink, being careful to work the ink film down to approximate printing thickness with a minimum of tapping. However, with this method, the inks will take longer to dry because they are usually too thick.

In any case, the proof or tap-out should be marked with the time it was made and inserted in a book or pile of paper or between glass plates. Then, every half hour, it should be removed and tested for dryness by rubbing a clean, dry finger across it.

Several instruments are available for testing ink drying time: the NPIRI drying time recorder, which measures the time required for a print to dry free of ruboff; the IGT drying time recorder, the Laray drying meter, and the Pira print drying tester, which measure the time required for a print to dry setoff free; the Pira ink drying tester, which determines the time for ink films to dry on a glass plate; and the Gardner drying time tester for impervious surfaces. These instruments are more precise than rubbing the film with a finger, but, because there are so many variables in ink drying, the accuracy with which they can predict the time required for a pile of prints to dry is probably no different.

Web inks. Drying time for web offset inks can be tested with a Sinvatrol tester. Developed by Sinclair and Valentine, the Sinvatrol tester measures conveyor speed and temperature, giving information used to predict specific dryer and web temperatures needed to dry heatset web offset inks on a given press. It is also useful in testing flexo and gravure systems.

A wet, printed sample is passed through the tester on a conveyor. Conveyor speed is variable between 0 and 100 ft./min. (0 and 0.5 m/sec.). Drying temperature is

Sinvatrol tester
Sinclair and
Valentine Co.

controlled, but variable between 100°F and 600°F (38°C
and 316°C). In the laboratory, the Sinvatrol tester has
been used by printers and ink manufacturers to develop
low-energy printing systems.

Flexographic and gravure inks dry directly in proportion
to the content of low-boiling solvents. To increase the speed
of drying, the printer can add a "fast" (low-boiling) solvent.
To decrease the speed, the printer adds a "slow" (high-
boiling) solvent.

Lack of a good press-side drying tester for these inks is of
concern to printers. It has been suggested that two inks
can be compared by drawing them down in a fineness-of-
grind gauge and comparing how fast they dry.

Fineness of Grind

Ink pigments must be well dispersed in order for the ink to
perform well on press. Poorly ground ink contributes to
piling on press rollers, plates, and blankets; weak colors;
streaking; plate fill-in; unnecessary wearing of plates,
rollers, and gravure cylinders; and scumming. Therefore,
the fineness-of-grind test (ASTM D-1316) is one of the most
important ink tests. It measures pigment dispersion and
indicates the presence or absence of particulate debris in
ink. The fineness-of-grind test is simple to perform and is
used routinely in ink manufacture to ensure that the ink
has been ground properly.

The fineness-of-grind gauge is a tool-steel block with a
highly finished surface. Two or three parallel channels that
are 0.001 in. (0.025 mm) deep at one end and tapering to
zero depth at the other have been ground into the surface.
Alongside each channel are marks indicating graduations
of 0.0001 in. (0.0025 mm) in depth. A doctor blade or
scraper is supplied with the instrument. In order to
perform the test:

1. Place a small quantity (about 0.5 ml) of ink across the
deep end of the channels using the tip of the spatula. Two
samples of the same ink can be used, or a test ink can be
compared with a standard ink.

2. Draw the inks down toward the shallow end of the
channels with the scraper. Use a smooth, steady, slow pull
and enough pressure to wipe the bearing surfaces of the
block clean. This stroke should take at least 4 sec. As the
depth of the channels decreases, any particles that will not
pass between the scraper and the channel bottom produce
readily visible streaks in the remaining ink films.

3. Find the point at which there are as many as four scratches. The higher the number on the block where the streaks start, the larger the size of the dirt, undispersed pigment, or grit. Then, find the point at which there are ten scratches. The greater the difference between these points, the broader the size range of the largest particles.

Evaluation of results obtained with the fineness-of-grind gauge is controversial. The test cannot predict that an ink will pile or cake, but when an ink piles, it can indicate whether the pigment was improperly dispersed.

Wet Ink Film Thickness

An ink film thickness gauge enables the printer to measure the amount of ink being carried by the printing press. As previously described, the thicker the ink film, the longer the setting time required. Furthermore, in magnetic ink printing, ink film thickness is critical because it determines the print's signal strength. An ink film thickness gauge is required for laboratory testing of inks where the thickness of the wet ink film is often critical to the results of the test. (See ASTM D-1212.)

The Interchemical ink film thickness gauge is essentially a roller with a center portion ground away to varying depths around the circumference. It is rolled over the ink surface until a point is reached where the ink just fails to touch the bottom of the recess. At this point, the corresponding number of the scale around the side of the gauge gives the thickness value.

As with all test instruments, a newly acquired ink film thickness gauge should be calibrated. If the instrument is faulty, it will give poor results. Unfortunately, new test instruments do not always give correct readings.

Interchemical (Inmont) wet film thickness gauge with holder
Courtesy Gardner Laboratory, Pacific Scientific Co.

Tack

The printer needs to control ink tack to prevent picking of the paper and to ensure proper trapping in multicolor printing. The ink manufacturer is primarily interested in tack as a measure of quality control; whether the ink under study has the required properties and whether it has the same properties as the last batch.

Printers used to judge tack with a finger tap-out (see Chapter 5, "Flow"). This outmoded test is highly subjective and nonquantitative. Results are not closely repeatable, even by an experienced observer. The GATF Inkometer and several other instruments give numerical results that are far more useful in measuring or comparing inks.

GATF Inkometer. Robert Reed of the Lithographic Technical Foundation, predecessor to GATF, developed the Inkometer in 1937 to overcome the difficulties of judging tack with a finger tap-out. The Inkometer enables standard values for tack to be reproduced within reasonable tolerances in successive batches of paste inks, and is used by most American ink manufacturers for this purpose.

The Inkometer is a machine that simulates the rollers on a press and measures the force required to split an ink film at press speed. There are three rollers: a driven, brass

GATF Inkometer

roller maintained accurately, usually at 90°F (32°C), a rubber vibrator that distributes the ink evenly, and a rubber rider roller that is pulled in the direction that the brass roller turns. A counterweight on a bar connected to the rider roller measures the force required to keep the rider roller in place. This force (or torque) is referred to as tack. Ink film thickness, speed, and temperature must be carefully controlled.

Both manual and electronic inkometers are available. The manual Inkometer operates at roller speeds of 400, 800, or 1,200 rpm, and 400, 1,200, or 2,000 rpm. The electronic model is continuously variable from 100 to 1,200 rpm, or from 100 to 3,000 rpm. It has a digital readout and is equipped with a plug that permits connecting it to a continuous chart recorder. Using a chart recorder, it is possible to follow instantaneous changes in readings and to obtain a graphic readout. The Inkometer speed ranges bracket most commercial printing speeds on modern sheetfed and web offset presses. The instrument works best when operated in a room maintained at a constant temperature.

GATF electronic Inkometer
Courtesy Thwing-Albert Instrument Co.

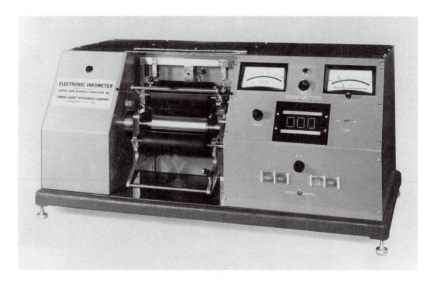

The geometry of the Inkometer is such that the line—linear speed—is related to the speed in revolutions per minute as shown in the accompanying table.

The tack of the ink should be measured at the speed most closely matching the speed of the press on which the ink will be run.

Inkometer speed	rpm	ft./min.	m/sec
	400	314	1.60
	800	628	3.19
	1,200	942	4.79
	2,000	1,570	7.98
	3,000	2,355	11.97

Because of instrumental variations, any two instruments that are to be used for communication purposes (as between two different plant locations or between customer and supplier) must be calibrated against each other. The electronic Inkometer can be zeroed in and is easier to adjust than the manual model.

Operating procedures. There is, unfortunately, no standard operating procedure for the Inkometer. Results vary, depending on the procedure used, and, therefore, when comparing inks, it is important to use precisely the same procedure for each ink.

GATF carries out the following procedure in a room maintained at $73.5\pm1\,°F$ ($23.0\pm0.5\,°C$) and at $50\pm2\%$ R.H. The brass roller is tempered with water at $90\,°F$ ($32\,°C$).

The ink pipette is used to deliver exactly 1.32 ml of ink. The rolls are turned by hand to start distributing the ink, then the machine is started at 400 rpm, and a stopwatch is started at the same time. The tack readings are taken every 30 sec. for the first 2 min., then the speed is increased to 800 rpm. The machine is permitted to run at 800 rpm for 20 sec. and another reading is taken. The speed is then increased to 1,200 rpm, allowed to run for 20 sec., and a reading is taken. Lastly, the speed is reduced to 400 rpm for the remainder of the 10-min. test. Web heatset and nonheatset inks are run at 1,200 rpm. The results show the tack of the ink at three speeds and the stability of the ink at $90\,°F$ ($32\,°C$) for 10 min.

After each use, the Inkometer rolls must be thoroughly cleaned with VM&P naphtha or a similar solvent. The rolls must be thoroughly dry before more ink can be applied. The VM&P naphtha can be wiped away with a cloth wet with alcohol to speed the drying process.

Other instruments that use a roller for measuring tack are manufactured by Prüfbau, IGT (Tack-o-Scope), Churchill, and Laray.

Viscosity

Viscosity is the resistance to flow, and the study of flow is called **rheology.** The viscosity of an ink determines how it flows on the press. It affects the tack, penetration, drying, gloss, rub resistance, dot gain, and color of the print. Instruments that measure viscosity, called **viscometers,** range from relatively simple viscosity cups to sophisticated electronic instruments. This chapter mentions viscometers that are commonly used for measuring the viscosity of printing inks.

Because viscosity is affected by changes in temperature, the temperature of the ink must be controlled carefully if the measurement is to be meaningful. Careful determination and interpretation of viscosity of printing inks is a research problem. However, simple devices are available that can be used for quality control or control of liquid inks on the press.

Cup viscometers. Cup viscometers (efflux cups) are commonly used to determine the viscosity of flexographic and gravure inks. These are cylindrical cups in which a capillary (Shell cup) or a machined hole (Zahn cup) controls the rate at which the liquid flows from the cup.

A stopwatch is used to measure the time (seconds) required for a full cup to empty. Recorded are the time, the cup number and the temperature at which the

Zahn viscosity cup
Courtesy Gardner
Laboratory, Pacific
Scientific Co.

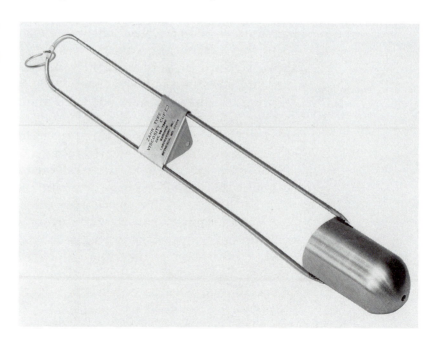

determination was made. The Gravure Association of America has standardized the Shell cup for rating the viscosity of gravure inks. ASTM D-1200 refers specifically to the Ford cup, but the procedure is applicable to Zahn and Shell cups.

Bar or rod viscometers. The Laray and Churchill falling-rod viscometers are useful for measuring the viscosity of paste inks for offset and letterpress printing. These viscometers measure the time required for a rod to fall through the liquid being tested. The viscosity is then determined by plotting the readings on a chart. The plot gives both the viscosity (or plastic viscosity) and the yield value.

Laray viscometer
Courtesy Testing Machines, Inc.

Brookfield dial-
reading viscometer
*Courtesy Brookfield
Engineering
Laboratories, Inc.*

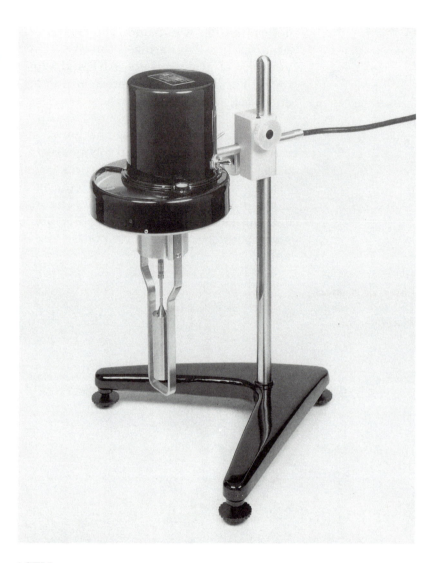

ASTM tests for
viscosity

ASTM Number	Brief Title
D-1200	*Viscosity of Paints, Varnishes, and Lacquers*
D-301	*Soluble Cellulose Nitrate*
D-1725	*Viscosity of Resin Solutions (Bubble Method)*
D-1545	*Viscosity of Transparent Liquids by Bubble Time Method*
D-2196	*Rheological Properties of Non-Newtonian Materials (Brookfield)*
D-445	*Kinematic Viscosity and Dynamic Viscosity*

Rotational viscometers. Rotational viscometers, like the Haake and Ferranti-Shirley, are expensive, complex, precision instruments for research, product development, and quality control of inks, varnishes, paper coatings, and other liquids. They are not the type of instruments to be found in most printing plants.

The Brookfield Synchro-Lectric viscometer consists of a spindle driven by an electric motor. It provides a numerical readout of different resistances to flow, but because the rate of shear is indeterminate, it cannot give a true viscosity reading. Since the design of this Brookfield viscometer permits continuous monitoring of viscosity, it is used to monitor the viscosity of flexo or gravure inks while they are on press.

Vibrating-reed viscometers. In these viscometers, the vibration of a reed is damped by a liquid, and the amount of damping is related to the resistance to flow (the viscosity) of the liquid. These viscometers, like the Brookfield, give continuous measurements of viscosity and are used to monitor the viscosity of flexo and gravure inks on the press.

In-process viscosity measurements. Much effort has been directed to developing devices for continuous in-process viscosity measurements. Flexographic and gravure inks are relatively fluid, with a high solvent content. The viscosity of solvent-based inks changes as the solvent evaporates. Water-based inks often change viscosity if there is any change in pH. Changes in viscosity obviously change the ink flow, and this affects print color and other properties. Instruments that monitor ink viscosity on the press and control the addition of solvent to maintain a uniform level of viscosity are available.

Flow

The flow of a small amount of ink under the force of gravity is related to the dispersion of the pigment. A test for relative flow properties of two or more inks is made with a flow plate, such as the Columbian Carbon flow plate. The test can be conducted as follows:

1. Thoroughly work the inks to be compared on the ink slab.

2. Place exactly 3.0 grams or 3.0 ml of each ink in the upper end of a channel and tilt the bed to the desired angle.

3. Record the length of flow after 15 min. The flow plate is helpful to ink manufacturers in testing the flow characteristics of dispersed pigments, especially blacks.

Length

Ink that is too short does not transfer properly (from roll to roll or from blanket to paper). Ink that is too long tends to fly or mist. While these problems are not common or serious, it is a good idea to check ink length.

Although there is no standard test to determine ink length, there are procedures to check it. To check length, place small, equal portions of the standard and test inks on the ink slab and work each ink thoroughly with a spatula to break down any false body. Compare the lengths of this by lifting several strings of each ink with a spatula or fingertip and estimating the distances at which the strings break.

Fly or Mist

Certain inks tend to "fly" or create fine mists of minute droplets upon transfer in fast-running presses. Small changes in ink formulation can influence fly. The Inkometer can be used to show the tendency of an ink to fly. It is often necessary to place a piece of white paper under the Inkometer rollers (behind them if the Inkometer is electronic) while running the tack test.

If a more rigorous test is required, perform the following test. Apply the ink to the Inkometer rollers using the ink pipette and allow it to distribute for 1 min. at the lowest speed. Then place an 8½×11-in. (215×280-mm) sheet of paper folded in thirds (with the fold up) underneath the metal and oscillating rollers for 1 min. Remove, unfold, and examine the sheet. If flying has occurred, the area not covered during the test will show a tint of the ink. If flying is not apparent, shift to the medium speed and repeat the test. If flying is still not apparent, shift to the high speed and repeat.

Flying or misting is common to letterpress inks where a fairly thick film (0.4 to 1.0 mil) is carried on the plate. It is rare with offset inks. When flying does occur with offset inks, it is caused by carrying too much ink on the press—usually a result of using ink with low color strength that requires a thick film to yield the proper color. Inks that emulsify excessive amounts of water can also cause some of the same problems.

Emulsification

There are many simple tests to determine the *tendency* of litho inks to emulsify. Unfortunately, none of them provides results that can be used to predict how an ink will perform on press. Inks that do not emulsify sufficient water and inks that emulsify too much water both perform poorly on the lithographic press.

The tendency of an ink to emulsify is best determined with a Duke tester, which controls the rate at which ink is mixed with water. Ink and water are mixed in the Duke tester for 10 min., and a sample is removed each minute. The water content in the emulsified ink sample is determined and a plot of water content vs. time is drawn. The accompanying figure illustrates both good and poor lithographic inks. The ink represented in curve A picks up too much water. It will tend to transfer to both image and nonimage areas, causing scumming, emulsification, low print density, and snowflaked solids. The ink on curve A will "gray out." The ink plotted on curve B can probably be run, but it will require constant attention. The ink represented in curve C, however, is an ideal lithographic ink. It will print sharply and be easy to run. Ink D cannot handle the water on press and the dots will tend to sharpen. Ink E cannot dispose of the water applied to the image area and will not print at all.

Surland water emulsification curves

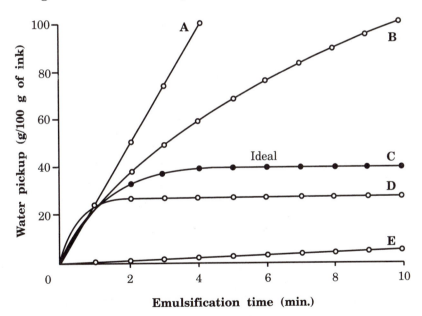

The GATF procedure for the Surland Emulsification Method using the Duke tester is as follows:

1. Place 50.0 g of ink in the bowl of the Duke tester. Record the weight of the bowl, ink, and beaters.

2. Set the Duke tester for 90 rpm.

3. Place 100 ml of distilled water or test dampening solution in a 250-ml beaker and note its pH and conductivity.

4. Add 15.0 ml of the water or test dampening solution to the bowl with the ink, and turn on the tester. It will shut itself off after 90 revolutions. (Note: web offset or news inks may require 30.0 ml of test solution.)

5. Lift the beaters from the bowl, pour the unemulsified water back into the beaker and weigh the bowl, mixing blades, and emulsified ink. The increase in weight is the amount of water emulsified. The amount of water emulsified times two is the percent emulsification.

6. Take another 15.0 ml of water from the beaker, add it to the emulsified ink and run the test for another minute (90 revolutions).

7. Repeat steps 5 and 6 until ten tests have been run (the ink has been emulsified for ten min.). Note the color, pH, and conductivity of the test solution, and draw a curve to show the amount of water taken up after each test. (The test takes about 45 min. to complete.)

Pigment Bleeding in Dampening Solutions

Since lithographic inks must come in contact with dampening solutions, the ink pigments must not dissolve or bleed in these solutions. (A test for pigment bleeding is described in ASTM D-279.)

The resistance of ink to bleeding can be judged quickly by placing a sample of ink in a mortar so that the bottom will be covered when the ink is spread out with the pestle. Add an equal volume of colorless dampening solution or a simulated dampening solution consisting of 0.2% gum arabic in water acidified to pH 4.5 with phosphoric acid. Work the ink and solution together with the pestle for 5 min. If alcohol is used in the fountain, it should be added to the test solution. With filter paper in the funnel, filter the solution. The extent of bleeding is judged by the color of the liquid passing through the filter paper. When the ink bleeds into the dampening solution, GATF retests it using distilled water. If the ink bleeds into distilled water, there is definitely a problem with the ink.

Scumming

GATF has developed a test that is useful for detecting the tendency of an ink to cause scumming. With a press that has a split fountain, the test can be used to compare two inks. The procedure is as follows:

1. Expose a presensitized (or other) lithographic plate to print a full solid, develop it, and put it on the press. Put the standard ink on one side of the ink fountain and the ink to be tested on the other side.

2. Run the plate on the press until a good overall impression is obtained. Use only plain water in the fountain.

3. While the press is running, lift the dampener rollers, allowing the plate to roll up solid and to dry.

4. Drop the dampening rollers to clean the plate. Run enough sheets through the press to show how well the plate has been cleaned.

5. Repeat the rolling up and the cleaning up several times until the differences between the inks (if there are any) have been established.

Density and Specific Gravity

Density is the weight/volume ratio and is measured in terms of the number of grams per milliliter. Specific gravity is the ratio of the weight of ink (or other substance) to the weight of an equal volume of water.

Inks are most commonly purchased on a weight basis but consumed on a volume basis. Therefore, it is desirable to determine the volume represented by a pound of ink, or the weight of ink per given volume.

Specific gravity of ink is related to ink mileage. In general, higher-gravity inks yield less mileage per pound than lower-gravity inks. Specific gravity and density are determined by NPIRI Method D-2 and by a number of tests listed in the accompanying table.

ASTM tests for density and specific gravity

ASTM Number	Brief Title
D-1217, D-1480	*Density and Specific Gravity of Liquids (Bingham Pycnometer)*
D-941, D-1481	*Density and Specific Gravity of Liquids (Lipkin Bicapillary Pycnometer)*
E-12	*Definition of Terms*
D-1475	*Density of Paint, Varnish, Lacquer*
D-153	*Specific Gravity of Pigments*

Standard weight-per-gallon cups are used in measuring such free-flowing materials as flexographic and gravure inks. After being filled, the cup is weighed. The weight of the ink in the container (in grams) divided by 10 is the

Weight-per-gallon
cups
*Courtesy Gardner
Laboratory, Pacific
Scientific Co.*

weight per gallon in pounds. The weight per gallon in
pounds multiplied by 0.12 is the specific gravity (1 gal. of
water weighs 1/0.12 or 8.3 lb.)

Flash Point

The flash point test is a measure of the flammability of an
ink or solvent. A sample is placed in a closed cup at a
temperature below its flash point. Then the solution is
slowly heated, and at regular intervals, the cup is opened
and a small flame is played over the surface of the liquid.
When the liquid reaches the flash point, the flame ignites
the vapors over the surface of the liquid. This temperature
is called the *flash point* because there is merely a flash of
fire over the surface of the liquid, and the liquid does not
continue to burn. When the temperature is raised further,
the fire or flame point is reached, at which the liquid
burns continuously. The test is described in detail in
ASTM D-1393 and E-502.

**End-Use
Properties**

End-use properties of inks affect the value or utility of
printed articles. Almost every printed package is designed
to sell merchandise of some sort. The printed film must
promote the product and its manufacturer or distributor.

Because of the wide variety of conditions to which packages, labels, and other products are subjected, a wide variety of tests have been devised. Many of them have not been standardized.

Some of the most widely used tests are described here. Further details are provided in the Methods of the Packaging Institute.

Abrasion Resistance

There are different types of damage or abrasion marks that occur after the product is shipped but before the customer receives it (Vandermeersche, *TAGA Proceedings,* 1988, pp. 395–412). Printers and ink manufacturers use terms such as scuffing, blotching, pick-off, ruboff, and scratches to describe this abrasion. Although it is influenced by shipping conditions and the properties of a particular ink, it is pretty much determined by the paper or other substrate. Ink is often not a factor (Tasker, *TAGA Proceedings,* 1986).

Abrasion resistance is important for magazine and catalog covers, book covers, labels, box wraps, folding cartons, and metal cans. The three instruments most commonly used to test for such resistance are the Gavarti Comprehensive Abrasion Tester (CAT), the Sutherland rub tester, and the Taber Abraser. Other abrasion testers include those by Gardner, Bendtsen, and Pira.

Gavarti Comprehensive Abrasion Tester. With the Gavarti CAT, materials to be tested can be cut to any size up to 4.5×4.5 in. (110×110 mm). The materials are placed in contact with each other and sandwiched between two synthetic plastic blocks. The surfaces of the two blocks are covered with a spongy layer that keeps the test samples in place during the rubbing action: no clamping, scoring, or folding is necessary.

A wide range of products can be tested: tissue paper, labels, magazine covers or inserts, folding cartons, book covers, corrugated boxes, plastic films or sheets, metal foil, or steel or aluminum sheets. Pressure is applied to the top and sides of the tester. The pressures constantly change during the test as a result of the motion of the carriage on which the blocks are resting. Both the frequency of vibration and the distance or span of the side-to-side movement can be set very accurately. The instrument is turned on for a set period of time, and at the end, the two surfaces are examined for wear or ruboff. The results are highly reproducible and linear. Vandermeersche uses a densitometer to determine the change in reflection density in order to quantify the results.

The Gavarti CAT correlates reasonably well with the Sutherland tester but very poorly with the Taber. Its main advantages over the Sutherland are its speed and reproducibility.

Sutherland rub tester. The Sutherland rub tester has been accepted as the standard means of testing abrasion by the National Folding Paper Box Association. The rubbing paper or board is scored (if necessary) and mounted snugly onto the rubbing block. There are two blocks: one weighs 2 lb. (0.9 kg), the other 4 lb. (1.8 kg). The printed sample for the test is mounted, printed side up, on the base of the tester and held in position by two pins. The rubbing block is attached to an arm that oscillates back and forth in an arc-like motion, rubbing the printed sample. The number of rubs is indicated by a counter dial. Paper cartons are commonly tested by giving fifty rubs with the 2-lb. block and eighty rubs with the 4-lb. block. Other types of packages have other requirements.

Two problems with the Sutherland are poor correlation between laboratories and very long running times for many new inks and coatings.

The Taber Abraser. The Taber Abraser has a rotating turntable with two revolving wheels. Various weights are applied to the wheels. A test sample is placed on the turntable and the wheels revolve over it. The results depend on the structure of the abrading wheels and the weight applied. Material abraded from the sample is measured with an analytical balance, or by suspending it in water and measuring the turbidity.

The Taber correlates poorly with field experience and with other test methods, partly because the abrasive wheel tends to clog with material from the sample.

Adhesion Tests

The adhesive strength of an ink to a substrate is frequently determined by hand-wrinkling, fingernail-scratching, "Scotch tape," and manual metal bending tests.

The **wrinkle test** is used only for liquid inks on flexible films. It is usually made by grasping a piece of the printed film between the thumb and forefinger of each hand and rotating the hands back and forth fairly rapidly, in opposite directions, about ten times. The wrinkled area is wiped free of any loose ink particles, and the printed surface is inspected for disruption.

To perform a **scratch test,** lay the sample film on a pad or cushion and quickly draw the back of the fingernail over the film without cutting it. The ink film is observed and the degree of scratching noted.

In the most common version of the **Scotch tape** test, 2 in. of a 4-in. (100-mm) strip of tape are firmly pressed down on the print. While holding the print down with one hand, quickly pull the free end of the tape upward, away from the pasted end. If the print is not ruptured by this lifting action, the adhesive rating of the ink film is 4. Adhesive values range from 4 down to 0, which describes complete removal of the printed ink from the substrate. Special adhesive tapes are sometimes used. This test is also used to determine adhesion of coatings and varnishes.

Manual bending tests on strips cut from printed metal sheets can be used for preliminary evaluation of adhesive qualities of coatings and inks. However, sheets must be put through actual forming operations to prove that the ink's flexibility and adhesion are satisfactory.

Closures, such as caps and crowns, are handled in hoppers during manufacture and application. This handling can cause scratching and marring of their printed surfaces. Therefore, inks and coatings used to decorate them must be highly scratch resistant. Scratch resistance can be tested in laboratory hoppers designed to duplicate production conditions.

Cans are often decorated after they are drawn. The inks, nevertheless, must withstand being jostled or abraded in tubs filled with ice to chill the beverage in the cans.

**Block
Resistance**

Blocking is the adhesion of printed products to each other. Block resistance is the ability of a printed foil or plastic film to withstand heat and pressure without sticking or transferring ink. If many tests are to be conducted, it may be wise to purchase a block-point tester. If only a few are required, the following test, which follows ASTM D-2793, will suffice.

1. Select printed samples that include a solid area that is at least 1.5 in. (38 mm) wide and 4.5 in. (115 mm) long.

2. Fold each print to one-third of its original size in a manner that provides face-to-face and face-to-back contact. The crease from one fold to another should be at least 1.5 in. wide. This will allow a weight of 1.5 lb. (0.68 kg) on a circular base of 1 sq. in. (6.3 mm²) to be uniformly distributed on the sample.

3. Place some rubber-covered glass plates in a 120°F (49°C) thermostatically controlled oven. This temperature is appropriate for most applications. (The test should be run under conditions that are slightly more severe than those encountered in actual practice to provide a margin for safety.)

4. Carefully stack the folded test samples on the rubber base and place the weight centrally upon the uppermost sample so that the weight base bears against the back of the top print. Include an unprinted piece of stock with the samples and expose everything under the described temperature and pressure conditions for 18 hr.

5. After the exposure, remove the samples from the oven and examine them for blocking.

If the printed sample is no stickier than the unprinted stock, give it a rating of 5. If the printed sample is markedly stickier than the blank, but does not transfer ink, rate the sample 4. With increasing degrees of ink transfer, rate the samples from 3 to 0, zero indicating complete ink film transfer. The drying history of the print and the relative humidity in the oven during the test are also very important and should be reported.

Blocking is also tested with the Koehler K5300 I.C. block tester, which consists of a number of calibrated springs that will exert pressure on the sample. GATF usually tests samples at 10 lb./in. and 500 lb./in. at the temperature of 100-120°F for one hour.

Skid Resistance

There are two types of friction—kinetic and static. Kinetic friction is equal to the force required to maintain movement, and static friction is equal to the force required to start movement.

Several methods have been developed for evaluating skid, or slip, resistance. To achieve quantitative, reproducible results, great care must be taken, paying attention to details. TAPPI Method 503 provides the coefficient for static friction of shipping bag papers and TAPPI Method 664 describes the kinetic friction of wax-coated papers.

ASTM D-1894 explains the measurement of static and kinetic friction.

Lightfastness   Resistance to fading is important if an ink is to be used on a poster, calendar, or other printed piece that will be exposed to light for a long time. Some pigments have much greater resistance to fading or discoloration than others. As a rule, inorganic pigments have better fade resistance than organic pigments, but because most inorganic pigments show weak, dirty colors, organic pigments are often chosen in spite of their tendency to fade.

The simplest method of testing inks for lightfastness is to expose normal prints to sunlight. In order to judge changes, part of the printed area should be masked during exposure. For indoor exposure, the prints should be about 1 ft. (0.3 m) inside a south window to allow direct exposure to sunlight and good air circulation. For outdoor exposure, they can be placed on an inclined board facing south. In some cases, the print may be protected from weather by window glass. Samples of the unprinted stock should be exposed in the same way and at the same time as the print. Any color change in the stock can seriously affect the color of the print, even if the ink pigment is lightfast. ASTM D-3424 describes a test for lightfastness.

The same pigment in various varnishes may exhibit different light-resistance properties. Relative humidity also affects lightfastness. Pigments in dilute dispersions (i.e., tints) behave differently from those at full strength. The light resistance of an ink is also influenced by ink film thickness.

The intensity of sunlight varies greatly throughout the year, and the most rapid fading occurs during the summer months. Exposure tests are best made during these months, usually for 30, 60, or 90 days. Latitude and altitude also affect the amount of sunlight falling on the sample: the sun is more intense in Miami than in Montreal, more intense in Denver than in Chicago.

Accelerated indoor fading tests are made with the Fade-Ometer, and outdoor exposure is simulated in the Weather-Ometer. A few hours in the Fade-Ometer is equivalent to many weeks of normal exposure. Furthermore, there are no rainy days to cloud the time of exposure. Fading due to light should not be confused with loss of color due to chemical changes such as poor resistance to alkalis.

Odor and Taste

Most common solvents are distinguished by their characteristic odors. Solvents used in inks usually have odors that cannot be tolerated in packaging. Inks that dry by oxidative polymerization may give off odors that smell like linseed oil. Packaging material for food is often required to have minimal or no odor. Testing the printed package for odor is very important.

A test for determining the odor of a printed ink film is described in ASTM E-462. It involves storing the sample in a sealed jar for 20 to 24 hours. The odor is observed by a panel of experienced judges who rate the odor from 0 (no odor) to 6 (strong odor). If a taint test is required, chocolate and unsalted butter are added to the jars and sampled for any imparted taste at the end of the test. A similar, but more involved method (described in ASTM E-619), is used for evaluating foreign odors in paper packaging.

Retained solvent can be determined with a gas chromatograph, used by some bakeries to test for potential odor problems. Judging taste and odor transfer from packaging film is described in ASTM E-462.

Heat-Seal Resistance

Inks used in packaging materials are often required to resist sealing damage. Resistance to sealing is measured by heat-sealing a printed sample and observing the print afterwards. For a sample to be heat-seal resistant, the ink must not melt, smear, or stick upon heat-sealing.

One test for heat-seal resistance uses the Sentinel heat sealer, which consists of heated jaws with a Teflon cover. The printed material is put between the jaws. The temperature, pressure, and swell time are adjusted, and a button is pressed to activate the machine. After the test, the sample is removed and examined.

Another test involves a temperature-controlled iron that is moved over the surface of the ink to see if the heat causes smearing. This simulates the heat applied on a side seam. Usually, the temperature at which the ink will start to soften is its maximum heat-resistance temperature.

Water Resistance

Three tests of water resistance are commonly used:
- Resistance to bleeding in water at room temperature.
- Adhesion when subjected to ice water or to freezing and thawing.
- Resistance to boiling water.

To test for bleeding in water at room temperature, place the printed film in contact with a damp filter paper or towel for 1-5 hours and note any bleed or color transfer.

The freeze-thaw test involves immersing the printed film in ice water for a time and examining for loss of adhesion, or freezing the film and then testing adhesion upon thawing. Loss of adhesion is often noted for a short time as the sample thaws, but once the temperature rises, adhesion is regained. During the transition period, ink sometimes transfers from one surface to another.

To examine resistance to boiling, place the package in boiling water and examine the adhesion of the ink. This test is important for inks and coatings on boil-in-bag packages.

Two other methods of determining boiling resistance are described in ASTM D-279.

Lamination Tests

When ink is laminated between a substrate and adhesive, it must adhere well to both. The bond strength of the laminate is the force required to pull the surfaces apart. It is measured on a tensile tester such as the Amthor. Care should be taken to ensure that inks for laminating do not exhibit excessive blocking tendencies in the rewind before lamination. This is not necessary if the lamination is performed in-line.

11 Specification of Printing Inks

Specifications are a means of communication between the customer and supplier and should be developed cooperatively. The customer should not develop specifications and impose them on the supplier, nor should the supplier establish them and hide them from the customer.

In the best customer/supplier relationships, the printer provides the ink manufacturer with sufficient information about a particular job and asks the ink manufacturer to supply an ink that will be satisfactory for the job. Given this ideal scenario, all sorts of problems can still arise. Perhaps the biggest ones come from disagreements about what is satisfactory and from a lack of concrete test results to assist in communication between customer and supplier.

In addition to definitions of satisfactory, the ink manufacturer encounters difficulties exclusive to attempting to furnish a suitable ink. For instance, the viscosity of varnishes varies from one order to the next, as do the properties of pigments and other materials; and suppliers often change their sources, formulas, or specifications, or go out of business. However, with good quality control in the ink plant, repeat batches of the same ink formula should not vary considerably. To keep things under control, ink manufacturers specify and test both raw materials and finished products.

Far too often, ink and other products are bought on price alone because the printer thinks that price signifies the value of the product and its suitability for the job. Printers, however, should not look for the lowest priced product; instead, they should purchase the product that produces the lowest manufacturing cost. Specifications help the customer judge the value of the product to match against its price.

As our statistical quality control (SQC) or statistical process control (SPC) procedures teach more and more about the urgency of starting with the right materials, it becomes increasingly apparent that specifications must be developed to assure that the printer gets a product that will perform properly in the printing plant, ink plant, or other operation.

Developing Specifications

The following discussion deals with the properties that are most likely to influence profits. Development of useful specifications takes time and effort. The cost of these specifications must be returned in increased productivity, reduced waste, and products that command a higher price in the marketplace. The accompanying table lists properties that should be considered in establishing ink specifications. In addition to these properties, lightfastness, gloss, toxicity, product resistance, rub resistance, and heat resistance are sometimes specified.

Ink specifications

	Sheet-fed	Heat-set	Non-heat-set	Flexo	Gravure	Screen
Color	X	X	X	X	X	X
Color strength	X	X	X	X	X	X
Drying time	X	O	O	O	O	X
Emulsification	X	X	X	O	O	O
Fineness of grind	X	X	X	X	X	X
Open time	X	O	O	O	O	X
Shelf life	X	X	X	X	X	X
Tack	X	X	X	O	O	X
Tack stability	X	X	X	O	O	O
Viscosity	X	X	X	X	X	X
Volatility	O	X	O	X	X	X
Yield value	X	X	X	O	O	O

X = This property should probably be specified.
O = This property is not likely to cause problems.

The table of specifications is only a summary or guide. It is not possible to compress all of the required information into a single "X" or "O," so there are certainly many exceptions. The table, however, should help printers decide where to start in establishing a profitable program of ink specifications.

Likewise, the following values of specified properties are representative, or typical, values. They should be adapted to meet the needs of the printer specifying the ink.

Color. Color is one of the most troublesome properties to specify. Colorimetric values or L,a,b values are usually the most useful numbers to use. Acceptable variation from the specifications must also be included. The tests are run on a dry, printed ink film of specified thickness.

Color strength. Color strength can be determined using a densitometer on a dry ink film. It may also be specified as the ink film thickness that must be carried on the press to produce the required optical density on the dry print. (See "Sample Specifications for Printing Inks" table in Chapter 5.) For practical purposes, color strength is measured with

h-out test. It should vary no more that ±5% from
o batch. Color strength is an important specification
e it determines the value of the ink.

g time on paper. Drying time on paper is tested by
ng down or rolling out a sample of the ink to be used.
rint can be tested with the analyst's finger to
nine how long it takes to dry, but such a simple test
en unsuitable for a formal specification. Because
ent drying time testers do not give the same results,
est must be identified in the specification. A drying
of 2–4 hr. is highly desirable and may be a suitable
to start developing specifications. Actually, there is
eed to specify a minimum drying time, since the
ifications should include an open time or drying time
nonporous substrate.

ulsification. The Surland test indicates a lithographic
s ability to accept limited amounts of water. Inks that
pt too little or too much water cause printing problems.
fact that there is considerable controversy about what
stitutes too much water pickup makes the decision
icult. Numbers should be worked out between the buyer
l seller, but they should probably include maximum and
nimum water content after 2 min. and 10 min.

neness of grind. Inks that are poorly ground cause
ate wear and flow problems. The fineness of grind is
sily measured on a fineness-of-grind, or Hegmann, gauge.
eetfed offset inks should produce no more than four
ratches at 0.4 mil (0.0004 in.) or ten scratches at 0.2 mil
.0002 in.).

pen time. Open time or drying time on glass or other
onporous substrate is specified to prevent the ink from
rying on the press. Anyone who has had ink dry up on a
heetfed press knows the importance of open time. Sheetfed
nks rolled out on a glass plate with a Quickpeek tester
hould remain open or tacky for at least 4 hr.

Shelf life. Sheetfed inks should remain in good condition
for one year if the can is unopened and stored under
conditions approaching room temperature. If the ink is not
usable at the end of a year, the printer should ask the ink

manufacturer if it is possible to rework it into a suitable ink. Flexo and gravure inks may settle out in a much shorter period of time so that redispersing can be a problem. The ink manufacturer should assist whenever necessary, but the printer can stay out of trouble with good inventory control.

Tack. Since absolute tack values vary significantly from one instrument to another, printers must designate one brand and type of instrument as the standard. For process color printing, GATF recommends tack-sequence inks, so that each color may have a different tack. Batch-to-batch variations should be kept to ± 0.5 units.

Tack stability. Tack changes while the ink is being run. It is probably helpful to specify the tack after running the instrument for 10 min.

Viscosity. This fundamental property should be specified on all inks. Because of the profound effect of temperature on viscosity, the temperature of the test must also be specified. In addition to a raw material specification, viscosity is an important control when running liquid (flexo or gravure) inks. The viscosity of paste inks is suitably determined with a Laray falling-rod viscometer. For fluid inks, the Shell cup gives much greater precision than the commonly used Zahn cup.

Volatility. The volatility of heatset web inks must be properly controlled so that they will remain open on the press, but dry quickly and completely in the dryer. On a multicolor press, the first-down ink is usually less volatile than the last one. The printer may specify the boiling range of the ink, but heatset oils are identified by number (e.g., Magiesol 47), and the boiling range is a published number.

Yield value. Yield value determines whether the inks will flow under the force of gravity, behavior that sometimes causes sheetfed inks to "hang back" in the ink fountain. The property is determined together with viscosity, on the Laray viscometer. Only sheetfed inks are likely to cause a problem, and the yield value is of importance only with these inks.

Other properties may be specified as the need arises. Lightfastness can be measured in a Fade-Ometer or a Weather-Ometer, and the instruments can be used for developing specifications. Gloss is sometimes an important ink property, but it is affected by so many variables that specifications require great care: substrate, method of printing, and angle of measurement should be specified. Toxicity should be adequately covered in Material Safety Data Sheets so that further specifications are unnecessary. For some printed products, heat resistance may be important. In no case should ink color be affected by normal operation of the dryer on press. Several rub testers are available for measuring rub resistance, and the test must be described in detail if rub resistance is to be included in ink specifications.

Checking Specifications

Specifications require that the purchased product be checked to see if it meets the established requirements. This is an additional cost that must be kept in mind. It is unlikely that every shipment must be analyzed thoroughly, but some ink testing is required to be sure that the specifications are serving the purpose of improving communications between supplier and customer and of assuring that inks will meet the requirements of the job. Some tests are inexpensive and can be run by most printers on a routine basis; others are expensive and can probably be purchased from a laboratory or consultant more economically than they can be carried out in the printing plant. The ink testing equipment is less costly than the services of the people who run the tests.

12 Lithographic Inks

Since letterpress and lithographic inking systems are similar, letterpress and lithographic inks are similar. The press rollers distribute these paste inks by breaking down their structure, which reduces their viscosity (or body) making them flow readily and delivering a more uniform ink film to the image area of the plate.

Because lithography applies a thinner ink film to the paper than letterpress does, litho inks must be more highly pigmented than letterpress inks.

In addition to the differences in color strength, litho and letterpress inks differ in their sensitivity to water. Although it has been stated that the principle behind lithography is that oil and water do not mix at all, in reality the ink manufacturer must formulate lithographic inks that absorb or pick up *some* water. Totally water-repellent inks do not work well in lithography. Ink cannot be made to print on a surface wet with water, and this is the mechanism by which water keeps the nonimage area of the offset plate clean. As long as there is a film of water on the nonimage area of the plate, the ink applied by the form rollers will not stick to that area of the plate, but, if the image area becomes wet with water (for example, if gum is deposited on the image), then the ink will fail to adhere to the image. This problem is called gum blinding. The image can be seen on the plate, but the plate does not print the image onto the substrate.

The Inking System

The purpose of the inking system of a lithographic offset press is to provide a uniform ink film to the image area on the printing plate. The inking system is not a mixing device. Inks must be properly mixed before they are put into the ink fountain of the press.

Lithographic inks have a much heavier body than flexo or gravure inks. When the ink is printed on paper or another substrate, it increases in body and begins to set.

The ink fountain blade should be properly set so that a thin film of ink is applied to the ink fountain roller. This roller must also be properly set so that it can work the ink in the fountain as well as permit easier control of the ink flow. Note: it is easier to adjust the sweep of the fountain roller than it is to adjust the doctor blade.

Ink fountain keys must be set to allow more ink to flow to the solid areas because more ink is required to print a solid than a halftone dot. If the same amount of ink is fed from the fountain, the ink film thickness of the solids will be less than the thickness of the halftones.

The discussion of Stefan's equation in Chapter 5 shows the great importance of controlling ink film thickness if good trapping is to be achieved. Variations in trap are a major cause of color variations in both lithographic and letterpress printing.

The accompanying tables list representative lithographic ink formulations and appropriate quality control tests.

Typical formulations of lithographic inks	**Quickset Sheetfed Offset**	**Heatset Web Offset**	**Nonheatset Web Offset**
	Parts	**Parts**	**Parts**
	20 Phthalo blue	14 Phthalo blue	14 Phthalo blue
	24 Rosin phenolic ester	10 Calcium carbonate	10 Hydrocarbon resin
	24 Heatset oil	35 Rosin/phenolic resin	64 Mineral oil
	6 Wax compound	41 Heatset oil	12 High-boiling aliphatic solvent
	2 Paste drier		
	100	100	100

Appropriate quality control tests for lithographic inks	Color	Color Strength	Opacity/Hiding Power
	Gloss	Emulsification	Density/Specific Gravity
	Drying	Bleeding into Dampening Solution	Flying/Misting
	Tack	Fineness of Grind	Lightfastness
	Length	Masstone and Undertone	Rub and Scuff Resistance
	Viscosity	Tinctorial Strength	Product Resistance
	Odor		

Since formulation of inks is really of secondary importance to the printer, the subject is presented only briefly here.

Sheetfed Inks   A typical quickset sheetfed lithographic ink contains pigment, quickset varnish, drier, wax compounds, and solvents.

These sheetfed inks dry primarily by a chemical reaction, a polymerization reaction initiated by oxygen and catalyzed by the drier.

Sheetfed lithographic inks have a higher tack than web inks. Because of the slower speeds of the sheetfed presses, a higher tack is necessary in order to print a sharp dot. The quickset qualities of the sheetfed ink are developed by using a quickset varnish. A quickset varnish consists of a resin, usually combined with a drying oil, dissolved in a low-viscosity, high-boiling hydrocarbon oil. If the ink is properly formulated, the hydrocarbon oils will be absorbed by the paper coating, causing the viscosity of the printed ink film to rise rapidly.

Quickset inks not only permit the printed sheets to be handled more quickly than they could be if printed by a nonquickset ink, but they also set rapidly enough to increase the tack of the printed ink between units on the press. If they set sufficiently between units, it is possible to print four-color process inks that all have the same tack rating. This means that the printer can use one set of process colors for any printing sequence. Sometimes better trap and color uniformity can be achieved if a series of tack-rated inks (i.e., highest tack on first unit, lowest tack on last unit) is used.

Heatset Web Offset Publication Inks

A heatset web offset publication ink contains pigments, heatset varnish, wax compounds, and solvents.

The primary drying process for these inks is evaporation as the web goes through the high-velocity hot-air, but absorption and oxidation may also be involved. Chill rollers, located after the dryer, are required to complete the drying process.

The choice of proper heatset solvents is critical. If the oil evaporates at too low a temperature, the ink will be unstable on the press. If it evaporates at too high a temperature, the ink will not dry properly.

Sometimes the hydrocarbon resins remain soluble in the dried ink film and can be smeared by fingers and normal skin oils. Therefore, drier may be added to these heatset inks to provide harder drying.

Web offset lithographic inks have a high solvent content. Their color value is less, and their ink film thickness is greater than that of sheetfed inks. Solvent is added to web inks to reduce tack. This makes them less viscous than sheetfed inks. At the high speeds of web presses, inks as tacky as most sheetfed inks would cause problems.

Nonheatset Web Offset Publication and News Inks

Nonheatset web publication and news inks contain pigments, varnishes, and mineral oil.

Because carbon black by itself has a brown shade (or cast), a reflex blue (or toner) is often added to give the black a bluer cast. The vehicle is made up of a high-viscosity oil and a low-viscosity oil, blended to give the desired body.

Nonheatset web offset inks are usually used on highly absorptive, uncoated, groundwood stock. Coated papers provide too much ink holdout and usually cannot be printed with nonheatset inks.

Low-Rub Inks   Rub-off is a major problem with news inks, but rub-off can be greatly reduced by adding litho varnish to the mineral oil. Note: the litho varnish is more expensive than the mineral oil; adding it increases the price. Specific formulations are confidential.

Low-Solvent, Low-Odor Inks   Much progress has been made in developing low-odor, lower-solvent inks. Inks containing 25% less solvent than those previously used have been produced. Furthermore, by carefully treating the solvent, it is possible to remove the components that carry the odor and the ones that are known to react with sunlight to create smog and other air contaminants. Nevertheless, complete recovery or incineration of ink solvent is usually the only way to comply fully with environmental laws.

Radiation Drying Systems   Inks formulated with certain acrylic oligomers (partially polymerized resins) can be dried almost instantaneously by ultraviolet or electron beam radiation. Ultraviolet radiation has barely enough energy to initiate the reaction and, therefore, expensive initiators must be incorporated into the ink. However, the electron beams are much more energetic. Both generate hazardous rays and ozone and, therefore, must be properly guarded and vented.

Radiation-curing inks are chemically incompatible with conventional drying inks. Poor adhesion often results if a UV- or EB-curing varnish is applied over a wet, sheetfed ink. If the printer wishes to cover wet conventional inks with a radiation-curing coating, a water-based coating or other primer should be applied first. Confusion can be avoided by discussing the problem with the ink manufacturer.

Infrared radiation does not cure inks. As commonly used, infrared units increase the setting speed of inks. When infrared radiation is used with quickset inks, the speed of the setting can be extremely fast. Printers may need less spray powder when infrared dryers are used. Too much heat is worse than no heat at all. In general, the temperature of the printed sheet should not exceed about 120°F (49°C). Quickset/infrared inks that allow the printer to "work-and-turn" the sheets in 15 min. have been developed. These inks also dry in about 1 hr.

Metallic Inks

"Gold" or "silver" inks are made from bronze or aluminum powders. Compared to the usual ink pigments, they are very coarse and produce inks that can be difficult to run. These powders are handled better by gravure, where the required pigment level is much lower. Silver inks are usually easier to run than gold inks. Piling on the blanket, the plate, and/or the rollers is a typical problem with both. However, if the metallic powders are ground too fine, they can lose their luster. Applying two coats of a very diluted pigmented ink, if possible, may alleviate the problem.

Another method for printing gold is with a bronzing machine. A "gold size" base ink is printed on the paper or board, and then the bronze "gold" powder is dusted on the wet base ink. Collecting and removing the unused powder is a messy job.

Metal-Decorating Inks

A metal-decorating ink contains the following materials:
- Pigment: carbon black and organic and inorganic pigments selected to withstand both the high temperatures during drying and the high moisture of the pasteurization processes
- Vehicle: oxidizing drying oil/resin combinations, complex cross-linking polyesters or other polymerizable materials, and ultraviolet-reactive varnishes
- Solvent: heatset oil, 440–600°F (225–315°C) used with some systems; high-boiling solvents are used in modern, polyester formulations; others may be solvent-free, 100% reactive
- Modifiers: wax compounds and lubricants to withstand tooling during forming and passage through fill lines
- Drying methods: oxidation accelerated by high-temperature baking, polymerization by baking, ultraviolet light

Metal products are printed by both the offset litho process and the offset letterpress, or letterset, process. Offset printing is used because the hard, incompressible nature of tinplate makes it impossible to print directly from a metal plate. The rubber blanket used in offset printing proved ideal in transferring the image from the printing plate to the metal surface. Sheets of metal are fed through the press, printed, and then passed through long, high-temperature ovens to bake the inks. Most metal boxes are printed this way.

Metal cans and other containers are subjected to vigorous treatment during the forming, soldering, and filling operations. Additionally, the contents of the can may need cooking or high-temperature pasteurization; thus, the inks need to be specially formulated to withstand the treatment.

The cost of metal decorating is so high that ultraviolet-curing inks are entirely practical in this application. In the older process, inks were dried on the flat tinplate by curing them in large, expensive ovens for 10–20 min. before the cans were formed. This produces a hard, glossy film, but the process is slow and costly.

The radiation-curing inks can produce a finished product in seconds or less. The adhesion is satisfactory for cans to be used for cosmetics or insecticides, but when resistance to scratching or abrasion is required, as on beverage cans, a baking cycle is used.

In the early 1960s, manufacturers began to use aluminum for beverage containers. The metal cannot be soldered. The base of the can is formed by punching out metal ingots. Later the top is crimped on, producing a two-piece can. These cans are referred to as drawn-and-wall-ironed, or DWI, cans.

Because the body of the can is preformed, it became necessary to print it "in the round." As a result, the mandrel press was developed. On this machine, the can is held on a mandrel and the image is transferred to it from a rubber blanket to offset the ink onto the can. After printing, the cans are conveyed on pins through a high-temperature oven to cure them.

Magnetic Inks   Magnetic inks are designed primarily for imprinting bank checks with the E 13 B style of characters adopted by the American Bankers Association to enable the checks to be sorted electronically.

The magnetic properties of the ink are derived from a special crystalline form of magnetic iron oxide. Proper performance requires careful and precise formulation of the ink. Magnetic inks can be formulated as conventional, heatset, or quickset lithographic or letterpress inks.

Inks for Printing Plastic or Plastic-Coated Paper

Inks for printing on plastic or plastic-coated paper are sometimes called **collodion** inks because they are used to print on nitrocellulose (collodion) coatings.

Plastic films are more easily printed by gravure or flexography than by lithography. Most plastic films not only lack the stiffness needed to feed through the press properly, they also do not absorb the water used in the lithographic process. Ironically, control of dampening when printing on film is even more critical than when printing on paper. Nevertheless, plastic films and, much more commonly, plastic-coated paper are printed by lithography. Plastic-coated paper or board is stiff enough to feed well on a sheetfed press.

Quickset inks, which depend on the absorbency of a paper coating for setting, are totally unsuited for printing on plastic, which is largely nonabsorbent. Special high-solids inks that dry quickly are generally used. Ultraviolet inks are also used in printing on films and foils.

Furthermore, the smooth, glossy surface of the film promotes setoff of the ink printed on its surface. Using large amounts of spray powder destroys the high gloss that can be achieved by printing on plastic. This problem has been eliminated by the use of in-line aqueous or UV coatings that dry hard and prevent setoff without using spray powder.

Inks and coatings used for printing on nonabsorptive surfaces must have the necessary surface tension to properly wet and adhere to the substrate.

The ink manufacturer must be consulted before the printer attempts to print on any unusual surfaces; this is especially true for plastic surfaces. The printer should be cautious about mixing standard quickset inks with plastic inks because some of the aliphatic solvents could cause severe local warping of various plastic stocks.

Driographic Inks

In the driographic printing process, the planographic or offset plate has a silicone-rubber coating in the nonimage area to which ink does not adhere. The plate's failure to release ink completely and cleanly results in tinting or toning (ink in the nonimage area). Control of ink tack and its change with temperature is critically important in formulating a driographic ink. This is a major obstacle to overcome, and the reason why formulating a more satisfactory driographic ink is difficult.

Overprint Varnishes

An overprint sheetfed offset varnish contains the following materials:
- Pigment: none; sometimes clay for viscosity control
- Varnish: linseed oil, tung oil alkyd, phenolic resin
- Drier: cobalt and manganese salts
- Solvent: hydrocarbon
- Modifier: wax

Ultraviolet varnishes contain a special radiation-curing varnish and a photoinitiator instead of a drier. Overprint varnishes are inks without any pigment. The ideal overprint varnish is colorless and transparent, yielding good gloss and scuff resistance when dry. Overprint varnishes must also be stable on the press, dry rapidly, and adhere well to the print.

Water-based varnishes may be applied by special coating units or on the fifth unit of a sheetfed press. The inking system is disconnected, and the varnish is applied to the blanket from a coater or sometimes from a specially designed dampening system. These varnishes consist of emulsions of acrylic resins. They form films very rapidly after they are applied to the print, and varnished pieces can be cut or folded a few hours after printing. In addition to protecting the print, water-based varnishes provide considerable gloss.

Varnishes for ultraviolet curing can be applied similarly and, when cured with an ultraviolet lamp, give a surface that is resistant to scuffing, water, and most solvents.

Problems with Lithographic Inks

The most aggravating problems in lithography are often interaction problems, which arise when there is more than one deficiency among paper, ink, press, and press operation factors. Consequently, adjusting the ink may solve the problem temporarily, only to have it pop up later as a "different" problem.

Even when the ink is not the primary cause of the problem, it is often adjusted because changing the paper or changing the form to be printed is less practical. For example, a picking problem caused primarily by printing solids with an ink that is too tacky on a marginally suitable paper will be treated by softening the ink rather than by changing the paper or replacing the solids with a 90% screen tint. Softening the ink may change the problem from picking to loss of details in the shadows, ink emulsification, or setoff.

GATF observes that the greatest number of ink problems in lithography are related to poor drying. This includes setoff, chalking, and poor rub resistance caused by ink that is not suited to the paper because: (1) the printer did not order a suitable ink or did not give the ink manufacturer the correct specifications, (2) the printer added something to the ink that caused the problem, (3) the specified ink does not perform as expected, or (4) the ink manufacturer did not provide the ink that was requested. Other factors, such as using old ink, wide variations in tack, printing without a standard color sequence or standard set of process colors, and ordering from too many suppliers also contribute to printing problems.

Many of these problems can be avoided if the printer establishes a good rapport with the ink manufacturer, inspects or tests all incoming materials (including ink), and consults the ink manufacturer before altering the ink in any way.

Ink Contamination

In handling ink, the press operator should take every precaution to avoid contamination. Special care should be taken to prevent the mixing of dried ink (ink skin) into the fountain ink. In removing the skin from a can or kit, first cut around the edge with an ink knife and remove the oiled paper cover. Then, scrape off any ink evenly from the surface rather than digging deeply into the can. When the required amount of ink has been taken out, smooth the surface of the ink remaining in the can with a broad knife and either replace the old oiled paper circle or apply a new one. Press it into uniform contact with the remaining ink to remove any air bubbles.

Be sure that the press is clean before inking up, especially if a light color or tint is to be run. Tints can be markedly altered in hue or cleanliness by traces of a previously run ink remaining on either the ink or the dampening rollers. To avoid this condition, ink up the press with a light-colored ink or a transparent white ink and run it for several minutes. Then, scrape some of the ink off one of the drums to see how much its color has changed. Wash up the press and repeat until the ink is no longer contaminated.

There are a number of suitable press washup solvents. Spirits or some naphthas can be used. Usually they are blended for best results and sold under proprietary names.

Strong solvents effectively remove ink but also swell and ultimately harden rollers and blankets; therefore they should never be used. Formulation of a good press washup solution requires considerable knowledge and experience, as does everything else. In addition to being a good ink solvent, the washup solution should also be nontoxic, nonflammable, and compatible with all other press materials. Petroleum solvents, which are commonly used, should have flash points of at least 100°F (38°C) in order to meet underwriters' requirements. There are also two-step washup systems that remove ink as well as residues of gum arabic and paper, which tend to cause glazing and stripping of rollers. Properly used, these and other proprietary washup solutions are more efficient cleaners than petroleum solvents.

Heatset Dryer Problems

Web offset dryers are frequently referred to as "ovens" (perhaps because, unfortunately, they are sometimes used to cook the paper). Running the dryer at too high a temperature burns up expensive fuel, weakens the paper, reduces gloss, and promotes blistering and fiber puffing. By removing the moisture from the paper, the dryer also promotes cracking at the fold and gusseting.

Gusseting is a waviness in books that results when a very dry web is folded and bound. When the book is exposed to higher humidity, the pages swell except at the binding where they are physically restrained. There is no real cure for gusseting, but if the web is moisturized or rehumidified before it is bound, the problem can be prevented. Grain direction should always be parallel to the spine to reduce this problem.

Therefore, it is clearly in the printer's interest to use ink that dries at the lowest temperature consistent with stability on the press and to run the dryer at the lowest practical temperature. An optical pyrometer should be used to monitor the temperature of the web. Thermocouples that measure the flame or air temperature do not give the temperature of the web.

Ink in the Nonimage Area

Ink in the nonimage area involves a variety of problems, some of which are easily confused.

Dot growth, slur, and doubling increase the area of paper that is printed, and, in the printing of process colors, change the color of the print. Except for dot growth resulting from applying too much ink or ink that is too soft, these are not normally considered to be ink problems.

Catch-up, tinting, toning, and scumming are ink/plate/dampening problems that can often be corrected on the press. Catch-up (or dry-up) is the name applied to ink appearing in the nonimage area because of insufficient dampening of the plate. Tinting or toning results from ink emulsified in the dampening solution, while scumming is a permanent image—usually small specks—in the nonimage area. To distinguish toning, scumming, and catch-up, the wet-thumb test is useful. To perform the "wet-thumb" test, the press operator inserts a finger in dampening solution, then rubs it across the nonimage area of the plate and observes how the ink behaves.

Ink in the nonimage area

Problem	Cause	Wet-Thumb Effect	Solution
Catch-up	Plate too dry	Removes ink	• Increase dampening
Scumming	Poor plate	Does not remove ink	• Increase dampening • Clean plate • Replace plate
Toning	Ink emulsification	Removes ink	• Decrease dampening • Avoid detergents • Change inks • Check roller pressure
Ink dot scum	Plate corrosion	Does not remove ink	• Dry plates rapidly

Scumming can arise from a number of sources: film and film processing, plates and plate processing, dampening solution, lighting, press adjustment, paper, or ink. Before blaming the ink, the various possibilities must be checked out. The problem becomes especially complex when two or three of the possible causes occur simultaneously.

Ink can cause scumming in many ways. Improper formulation can create a scummy plate. Uncontrolled or unexpected changes in the pigment or the resin used by the ink manufacturer can also cause printing problems. Improper grinding that leaves grit in the ink accelerates plate wear and results in scumming and/or plate blinding. A test for scumming is described in Chapter 10.

Ink/Paper Problems

Application of a thin film of tacky ink, as in lithography, puts maximum stress on the paper. The paper may be damaged by application of such force, leading to curl and press problems such as piling, picking, linting, dusting, etc., which may be solved by the following:

- Changing the press settings
- Changing the ink
- Changing the paper
- Changing the blanket
- Changing the form

Adjusting the press (speed, blanket, pressure) is often the easiest way of solving these problems; changing the form to be printed is least likely to be an acceptable solution.

Solutions to piling, picking, linting, and curl are often applicable to other ink/paper problems.

Piling. Piling is the accumulation of material (usually dried ink or paper coating) onto the blanket or plate. It is extremely troublesome and complex. The major cause of image area piling is short ink caused by excessive emulsification with water and/or mixing with dust, coating, or other debris from the paper. Doctoring the ink with grease, shortening, or dry material, such as talc, can cause piling and should never be done.

As an ink becomes emulsified it becomes short. Short ink does not transfer properly. Imagine a small piece of putty (putty is very short) on an ink roller on the press. No amount of working on the press would cause it to flow evenly and transfer to the plate and blanket. Poorly ground or gritty ink tends to be short. Fibers, coating, or dust from the paper aggravates the problem. Water is mixed into the ink at the form rollers, while paper dust and debris may be added at the blanket, and it is at the blanket that piling is most likely to cause problems.

In addition to ink and paper, the blanket plays an important role in piling. If the surface of the blanket is very smooth, it forms a tight bond with the paper, pulling dust, fibers, coating, and pigment from the surface. Changing to a rougher blanket may help, but addition of a nonpiling additive (a glycol or a wax emulsion) to the dampening solution will also improve blanket lubricity and often solves the piling problem. Dampening solution additives, of course, should be purchased from the manufacturer of the dampening solution.

The accompanying table shows the many factors that contribute to piling. At first glance, it may seem impossible to print a job without serious piling, but after further consideration it becomes apparent that several factors may be adjusted or modified to cure a piling problem without contributing to other problems.

Sources of piling

Ink
- Poor selection of varnish promotes emulsification.
- Poor grinding results in gritty pigment.
- Excessive tack picks or causes linting or dusting of paper.
- Improper formulation promotes emulsification.
- Excessive pigment promotes shortness (improper pigment/vehicle concentration).
- Loss of ink solvent increases tack.
- Improperly dispersed pigment results in an ink that transfers poorly.

Paper
- Dusty or linty paper adds solids to ink.
- Insufficient binder or solubilization of binder causes paper coating to contaminate the ink.
- Low surface strength contributes to picking.

Dampening Solution
- Detergent or soap promotes ink emulsification.
- Excessive gum makes blankets sticky, promoting linting or dusting.
- Alcohol can promote precipitation of gum, which causes a sticky buildup on blanket.

Blanket Wash
- Excess detergent works its way into dampening solution, emulsifying the ink.
- Aggressive solvent creates tacky blanket that attacks paper.

Blanket
- Tacky blanket pulls materials out of the paper.
- Excessively smooth blanket pulls dust out of the paper.
- Excessive squeeze increases force on the paper surface.

Press
- Increasing press speed increases force on the paper surface.
- Increasing impression cylinder pressure increases dusting of paper.
- Reducing ink film thickness increases force on the paper surface.
- Excessive frictional heat on roller system evaporates ink solvents, increasing ink tack.
- Low temperature increases tackiness of ink.

Linting. Linting is an undesired pattern in the printing caused by accumulation of lint on the offset blanket. It is a major problem with newsprint and other uncoated groundwood papers. Removal of lint from the surface of uncoated paper is generally considered to be a paper problem, but it can often be solved by manipulating the ink or press. Decreasing the ink tack and increasing the flow of dampening solution are two ways to reduce linting. Increasing the squeeze or printing impression often reduces linting by scrubbing the accumulating lint from the blanket (even though it probably increases removal of lint from the paper surface).

Emulsification   As previously discussed, a good litho ink must emulsify some water but not too much. Even if the ink has the proper emulsification characteristics, it can become waterlogged by poorly formulated dampening solution, a press wash that contaminates dampening solution, a press wash that leaves detergent on the blanket, running forms with very low coverage, and running too much water. If the printer buys an inexpensive ink that has low color strength or color value, the press operator must run more of it to get up to color. The heavy ink film on the plate requires more dampening, and this can lead to excessive emulsification.

Slow Drying   Too much acid in the dampening solution and too much water on press can destroy or reduce the effectiveness of the drier in sheetfed offset inks. Excess acidity can also cause emulsification of ink in water, tinting, snowflaked solids, plate blinding, and dot sharpening or loss of detail in both sheetfed and web printing.

If relatively little water is emulsified in the ink and the stock is reasonably dry, the ink will dry properly even if the dampening solution pH is as low as 3.5. Thus, it is always important (and imperative when printing on nonabsorptive surfaces) to run the press with the minimum amount of dampening solution needed to keep the plate clean.

There is a strong correlation between the frequency of sheetfed drying problems and the pH of the dampening solution. Often, driers are not effective when the pH is too low. Also, some inks will break down and emulsify in the dampening solution when the pH is too low. Thus, a pH below 3.5 is never justified, although many inks will perform satisfactorily at a pH of 3.5–5.0. With acid dampening solutions, it is important for printers to follow the recommendations of their dampening solution manufacturers while being sure that the inks being used are fully compatible with the solutions.

Trapping

Trapping is the efficient transfer of ink from an offset blanket to a printed ink film. If the printed ink film is dry, it is referred to as **dry trapping.** On a multicolor press, ink must transfer to a wet film.

Differences in trapping are a major cause of color variation in letterpress and lithography. Causes of trapping problems with wet ink are entirely different from causes of trapping problems on dry ink. Trapping was previously discussed in Chapter 5.

Wet trap. To assure good wet trap, ink tack and ink film thickness must be kept under control. GATF has always recommended that inks vary by one or two tack units on multiunit presses, with the highest tack on the first printing unit. A thin ink film will not trap properly over a thick one; even if ink tack is properly controlled, ink film thickness must still be regulated. Therefore, ink film thickness increases gradually from the first to the last printing unit. Once the inks are formulated as to tack and strength, the printer should not vary the sequence on press in order to ensure the best trap achievable.

Since the optical density of a solid area is usually used as a checkpoint, color strength of the ink must be carefully controlled. If the color value of the ink is too weak, the system is thrown out of control.

Sheetfed printers often establish an internal standard much like the one in the accompanying table. The high ink film thickness of the yellow also improves gloss. On the negative side, high thickness promotes setoff and dot gain.

Example of tack and ink film thickness (four-color sheetfed offset press)

Color	Tack*	Ink Film Thickness†
Black	17	0.20
Cyan	16	0.28
Magenta	15	0.34
Yellow	13	0.40

*Inkometer reading at 1,200 rpm
†In mils, or thousandths of an inch

Ink manufacturers often offer process inks all of the same tack. These can be made to work satisfactorily if ink film thickness and ink setting are properly controlled, but even better results are achieved if the recommended tack sequence is observed, especially when printing on papers with high holdout. The ink manufacturer can then help customers by controlling the pigmentation level of the ink so that the printer must use the proper ink film thickness on press. Printers should specify inks of controlled color strength. On relatively absorptive sheets, tack and ink film thickness are not as critical for good trap.

Dry trap. Some inks, notably the quick-, hard-drying inks based on chinawood or tung oil, dry to form a hard, impervious surface that will not trap uniformly or accept other inks. This problem is commonly called **crystallization.** High drier content and IR drying are also believed to promote crystallization.

Hard waxes (for example, Teflon-type waxes), which give a scratch- and abrasion-resistant surface to the dried ink, also interfere with dry trapping; thus, the ink manufacturer usually avoids using them in inks that are to be dry-overprinted.

If excessive drier is added to the ink, the nondrying oil in which the drier is dissolved can rise to the surface and produce nonuniform areas on the dried ink, over which the subsequent inks will not trap.

Hickeys

Hickeys (small solid areas, sharply defined and surrounded by white halos) are sometimes referred to as doughnut or

Ink skin hickey

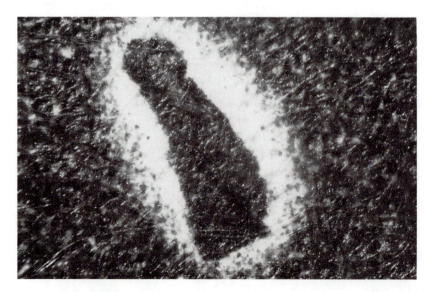

ink skin hickeys, but any source of dirt (the press, the pressroom, raw materials, crew) can cause hickeys. In addition to common sources of dirt, there are many unusual sources, and the solution to a hickey problem often involves a careful, lengthy search. Examining the hickey can give a clue to its source. Hickeys with dense, solid centers are caused by ink-receptive dirt. Void hickeys (hickeys without a visible center) result from water-receptive dirt, and hickeys with a partial or snowflaked center are caused by dirt that can be both ink- and water-receptive. Paint, spray powder, and other materials falling from the ceiling frequently cause hickeys.

Gloss Ghosting

Gloss ghosting is characterized by excessive gloss on the second side of a printed sheet, corresponding to the image on the first side. It occurs with oxidative-drying inks when the first side of the sheet is printed and there is not enough oxygen in the stack to dissipate the vapors yielded by the oxidizing reaction. The vapors, in turn, leave an invisible film on the back of the adjacent sheet. When this invisible film is printed over, the effect is much like the one achieved when printing over an overprint varnish. Gloss ghosting does not occur in heatset web printing.

Gloss ghosting is not an immediate reaction. It takes between four and six hours, perhaps longer, to occur. However, if the stack is exposed to air about two hours after the first side is printed, the possibility of gloss ghosting is often eliminated.

Gloss ghosting occurs more often on longer runs where the first side is printed and sits for a long time without being exposed to air. The longer it takes the ink to dry, the greater the likelihood of gloss ghosting occurring. In some cases, it can even result in fuming. Although lack of oxygen has been cited as the major cause of gloss ghosting, the paper, ink, dampening solution, printing sequence, and level of gloss also contribute to the reaction.

Gloss ghosting is not seen much any more because inks manufactured today have more solvent and a lower percentage of oxidizing materials. When it does occur, it helps to run the side with heavy coverage first if possible, and then, if time permits, wind the sheets about two hours after they come off the press. Another alternative is to wind the job after the first side is printed and, if there is enough time, allow it to set overnight on the chance that the vapors may evaporate. If neither of these is feasible, or do not work, call the ink company, explain what type of ghost is occurring, and send them sheets so that proofs can be pulled. The ink company can then supply a special varnish.

Mechanical Ghosting

Mechanical ghosting includes ink-starvation ghosting and "repeat" ghosting. The ghost image is always carried on the same side of the sheet. Mechanical ghosting results from inadequacies found in lithographic and letterpress inking systems where the press will not print ink uniformly when there is nonuniform coverage on the form. It does not occur in gravure or flexography.

When ink is removed from a form roller by a heavy form on the plate, the ink is not completely replaced by the ink splitting between the ink form roller and the vibrator roller or oscillator. (See Chapter 1 for a diagram of the inking system of the lithographic press.) Differences in ink film thickness on the roller result in differences in ink film thickness on the paper and cause color differences.

Consideration of the effect of ink film thickness on optical density suggests several steps that can be taken to reduce the problem.

Suppose that the variation in ink film thickness (IFT) on the ink form roller causes variations in IFT on the paper

Effect of ink film thickness on optical density

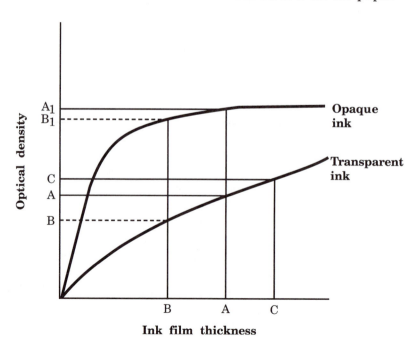

Ink film thickness

or that IFT A corresponds to OD (optical density) A and IFT B corresponds to OD B. By going to an opaque ink, differences in optical density are decreased from the difference between A and B to the difference between A_1 and B_1. By reducing the color value of the ink and running a thicker ink film, differences in ink film thickness are reduced, but running the ink at a higher ink film thickness (and a greater optical density) also reduces the difference in optical density, as shown in the illustration.

Emulsification of the ink with water is equivalent to printing a thinner film of ink, which aggravates ghosting. Use of alcohol or an alcohol substitute in the dampening solution is helpful.

Ghost areas that run around the cylinder can often be reduced by increasing the pitch (or sweep) of the oscillator. If mechanical ghosting is caused by improper diameter of the form rollers, these should be replaced by rollers of the diameter specified by the press manufacturer. A press with four or five form rollers is capable of better ink distribution than one with two or three. Some printing characteristics are already established when the press is purchased.

The following procedures may enable printers to overcome mechanical ghosting.
- Reduce water (add alcohol or wetting agent).
- Check diameter and hardness of form rollers.
- Be sure all ink rollers are adjusted and operating properly.
- Increase ink film thickness.
- Increase pitch of the oscillator.
- Use an oscillating form roller.
- Skew the image.
- Print ink take-off bars in the nonimage area.
- Use an opaque ink.
- Go to another press.

Piling Ghosting

A ghost of an image sometimes appears on the reverse side of coated paper printed on blanket-to-blanket web offset presses. This results from uneven pressure on the print caused by piling on the blanket adjacent to the image. If this blanket grows thicker because of piling, the added pressure will cause dot growth on the opposite side of the sheet, leaving an image. The cure for this ghosting is the same as for piling.

Mottle

Mottle is irregular and unwanted variation in color or gloss. It is often caused by rough and uneven paper coating. Sometimes, the ink manufacturer can increase the viscosity of an ink or formulate a semiopaque ink to counteract the effect of rough or uneven paper coating.

In general, the heavier the paper, the greater the variation in absorbency. Mottle is seldom found on uncoated book paper, only occasionally on cover stock, and frequently on carton board.

Variations in binder migration in paper coatings sometimes create mottle on coated papers. Like the ink, the binder detects differences in paper formation, or absorbency, and exaggerates them in the base stock.

Backtrap Mottle

As the number of printing units on a sheetfed press increases, the ink has more and more time to set before the last color is printed. Backtrap mottle is caused by partial, differential setting of ink on a multicolor press. Back trapping, the transfer of ink from the print to the blanket, is a common phenomenon. As long as the next sheet goes through in register, it usually causes no problem—the ink transferred from the print to the blanket is fed back to the next print.

However, if the ink has set unevenly by the time the print goes through the last unit, it may backtrap unevenly and create an uneven transfer to the next print. Backtrap mottle can be cured by:
• Changing the paper
• Changing the print color sequence
• Reducing or eliminating the quicksetting characteristic of the ink
• Reducing the number of colors being printed: printing several colors the first time through the press, letting the print dry, then running the rest of the colors

If the absorbency of the paper and the viscosity of the first-down ink are such that the ink starts to set while the paper is still on the press, backtrap mottle can occur. Changing either the paper or the ink will probably help. Changing the print color sequence will also change the amount of time that the ink causing the mottle has to set. Backtrap mottle is seen most commonly in cyan perhaps because cyan is usually the first color down and perhaps because the cyan pigments aggravate backtrap mottle.

Fluting

Fluting is the name given to the ridges or corrugations that run in the web direction on a heatset web offset press. As with ghosting and backtrap mottle, fluting is a very common problem.

The lower the basis weight of the paper, the greater the fluting. The problem is almost impossible to avoid on lightweight coated papers. Fluting is reduced by running the tension on the press at the lowest possible level. The lower the level of dampening, the less the fluting. Therefore, inks that require a low level of dampening should be chosen. Low-boiling inks are helpful because high temperatures in the dryer aggravate fluting.

Troubleshooting Lithographic Ink Drying Problems: Sheetfed

Problem	Cause	Solution
Blocking: printed sheets stick together.	See the problem "Setoff and blocking" in the table "Troubleshooting Lithographic Ink Problems: Ink in Nonimage Area."	
Chalking: pigment is dry and chalky.	Paper is too absorptive.	• Get ink better suited to paper. • Add heavy, water-resistant varnish.
	Ink is too soft.	• Get ink better suited to paper. • Add heavy binding varnish.
	Ink contains too much pigment.	• Get ink better suited to paper. • Add varnish and drier.
	Ink dries too slowly.	• Add drier. • Overprint with hard varnish. • See the problem "Ink dries too slowly."
Gloss ghosting: ghost of image from reverse side of sheet appears.	Ink dries too fast.	• Get ink better suited to paper.
	Fumes from drying ink alter drying on backup side.	• Wind the load. • Stack small lifts. • Print side with heavier ink coverage first. • Allow first side to dry 24 hr. before backing up. • Use slip sheets between prints.
Ink dries too slowly: ink should dry in 2–4 hr.	Ink contains insufficient drier.	• Check ink on stock before going to press. • Add drier. • Have ink manufacturer reformulate ink.

Troubleshooting Lithographic Ink Drying Problems: Sheetfed

Problem	Cause	Solution
Ink dries too slowly: ink should dry in 2–4 hr. *(continued)*.	Ink has too much drier (excess drier plasticizes the ink).	• Replace with ink containing less drier.
	Ink is too soft.	• Add body gum or replace with a stiffer ink.
	Dampening solution is too acidic.	• Get better dampening solution concentrate. • Keep pH at 4.5 to 5.5.
	Insufficient air is available.	• Wind the load. • Stack small lifts. • Increase antisetoff spray.
	Relative humidity is too high.	• Use dehumidifier. • Air-condition the plant. • Keep water in ink to a minimum (use alcohol). • Wind the load. • Add drying stimulator to the dampening solution.
	Ink is waterlogged.	• Replace with a more water-resistant ink. • Add body gum to ink. • Set rollers and press properly. • Add alcohol or a alcohol substitute to the dampening solution.
	Paper is too acidic.	• Replace paper.
Insufficient rub resistance: dried prints smear when rubbed.	Ink dries too slowly.	• See the problem "Ink dries too slowly." • Have ink manufacturer reformulate ink.
	Rub resistance is poor.	• Add wax to the ink.
	Ink is too soft.	• Have ink manufacturer reformulate ink. • Add heavy binding varnish.

Troubleshooting Lithographic Ink Drying Problems: Sheetfed

Problem	Cause	Solution
Insufficient rub resistance: dried prints smear when rubbed *(continued)*.	Paper is rough and abrasive.	• Get ink better suited to paper. • Add appropriate wax to ink. • Varnish the prints. • Change paper.
Scuffing: overly hard dried film abrades in folder or on handling.	Ink has dried too hard.	• Add wax. • Reduce drier in ink. • Change stock.

Troubleshooting Lithographic Ink Drying Problems: Web

Problem	Cause	Solution
Ink dries too fast: ink is unstable on press.	Ink is too volatile.	• Have ink manufacturer reformulate ink.
Marking: scratches appear on print.	Ink dries too slowly.	• Have ink manufacturer reformulate ink. • Use low-heat ink.
	Dryer temperature is too low.	• Increase dryer temperature.
	Web is too hot.	• Check chill rolls. • Control temperature with pyrometer.
	Air bar on turn bar is abrasive.	• Apply release agent to bar.
Printed sheets are sticky.	Ink dries too slowly.	• Have ink manufacturer reformulate ink. • Use low-heat inks. • Increase dryer temperature. • Control temperature with pyrometer.

Troubleshooting Lithographic Ink Problems:
Ink in Nonimage Area

Problem	Cause	Solution
Dry-up (catch-up, dry-back): ink is easily washed from nonimage area, but there is no ink color in dampening solution.	Plate is too dry.	• Increase water feed.
	Plate carries excess ink.	• Decrease ink.
	Dampeners are worn.	• Replace dampeners.
Ink dot scum (oxidation) not caused by ink: ink gathers in holes burned into plate by dampening solution.	Dampening solution is corrosive.	• Get better dampening solution concentrate. • Keep pH at 4.5 to 5.5.
	Dampening solution evaporates slowly from plate.	• Raise dampeners before stopping the press • Dry plates rapidly when stopping the press.
Ink flying or misting: strings or droplets of ink are thrown from the rollers.	Too much ink on rollers.	• Cut back on ink feed.
	Ink color strength is too low.	• Get a better ink.
Plugged screens or reverses, slur, dot gain: small areas to be left unprinted are filled with ink, halftone dots are too large.	Ink is too soft.	• Replace with a stiffer ink. • Add heavy binding varnish.
	Too much ink is on the plate.	• Reduce ink (and water) feed. • Replace with ink of higher color value.
	Dampeners are dirty or worn.	• Clean or replace dampeners.
	Dampeners are out of adjustment.	• Adjust dampeners correctly.
	Dampening solution is too weak.	• Add gum or get another dampening solution.
	Ink has been poorly ground.	• Have manufacturer regrind ink.

Troubleshooting Lithographic Ink Problems:
Ink in Nonimage Area

Problem	Cause	Solution
Plugged screens or reverses, slur, dot gain: small areas to be left unprinted are filled with ink, halftone dots are too large *(continued)*.	Ink rollers are improperly adjusted.	• Set form and distributor rollers correctly.
	Ink is too hot.	• Air-condition pressroom. • Adjust or repair ink rollers.
Scumming: ink cannot be washed from nonimage area.	Lith film may be to blame.	• Check film and film processing. • Reshoot film.
	Litho plates may be to blame.	• Check plates and plate processing. • Regum or remake plates. • Be sure plates are protected from light prior to exposure and developing.
	Worn-out plates may be to blame.	• Replace plates. • Control plate/blanket pressure. • Check roller pressure.
	Paper may be to blame.	• Replace paper.
	Dampening solution is too weak or too acidic.	• Add gum or replace with an alternate dampening solution concentrate.
	Ink is too water-resistant.	• Have ink manufacturer reformulate ink.
	Ink contains free acids.	• Replace the ink.
	Pigment dispersion is poor.	• Have ink manufacturer regrind ink.
	Ink is too soft.	• Replace ink. • Add heavy varnish.
	Dampener rollers are worn.	• Replace dampener rollers.

Troubleshooting Lithographic Ink Problems:
Ink in Nonimage Area

Problem	Cause	Solution
Scumming: ink cannot be washed from nonimage area *(continued)*.	Dampener rollers are poorly adjusted.	• Adjust dampener rollers.
	Form rollers are poorly adjusted.	• Adjust form rollers.
Setoff and blocking: printed ink film transfers to back of next sheet. Blocking results when the two sheets are stuck together.	Inks sets too slowly.	• Replace with quicker-setting ink. • Apply infrared heater.
	Ink film is too thick.	• Reduce ink feed. • Get stronger ink if necessary. • Use undercolor removal in area of overprinting.
	Ink is too soft.	• Choose proper ink. • Add body gum or heavy binding varnish.
	Paper holds out ink excessively.	• Get ink better suited to paper. • Apply infrared heater. • Change paper. • Stack small lifts.
	Spray powder is insufficient.	• Increase powder application. • Use a coarser spray powder.
	Static presses sheets together.	• Use static eliminator.
	Prints in load are overheating.	• Get ink better suited to paper. • Keep drier to a minimum. • Wind the load. • Stack small lifts.
	Ink has too much drier.	• Have ink manufacturer reformulate ink.

Troubleshooting Lithographic Ink Problems: Ink in Nonimage Area

Problem	Cause	Solution
Tinting (toning): color appears on plate and in dampening solution.	Ink emulsifies in water.	• Check and adjust roller settings. • Reduce water feed (add alcohol). • Add water-resistant varnish.
	Detergent or soap in dampening solution.	• Eliminate use of soap or detergent. • Replace soap with approved wetting agent.
	Pigment bleeds into dampening solution.	• Have ink manufacturer reformulate ink.
	Ink is not water-resistant.	• Have ink manufacturer reformulate ink.
	Pigment is poorly dispersed.	• Have ink manufacturer regrind ink.
	Ink is too soft.	• Replace with a stiffer ink. • Add body varnish.
	Overheated ink rollers emulsify the ink.	• Adjust or repair rollers.

Troubleshooting Lithographic Ink Problems:
Insufficient Ink in Image Area

Problem	Cause	Solution
Emulsification or snowflaking: water emulsifies into ink, producing weak color and unprinted spots.	Ink is not water-resistant.	• Have ink manufacturer reformulate ink.
	Ink and water rollers are out of adjustment.	• Check and adjust roller settings.
	Ink is too soft.	• Replace with a stiffer ink. • Add body gum or heavy binding varnish.
	Dampening solution contains detergent or soap.	• Eliminate use of soap or detergent in dampening solution. • Replace soap with approved wetting agent.
	Ink color is too weak. Low color strength requires heavy use of ink and water.	• Buy better ink. • Have ink manufacturer reformulate ink.
Insufficient color strength: print color is too weak.	Water pickup is excessive.	• Have ink manufacturer reformulate ink. • Avoid use of soaps. • Check press settings. • See the problem "Tinting" in the table "Troubleshooting Lithographic Ink: Ink in Nonimage Area."
	Ink transfer is poor.	• Have ink manufacturer reformulate ink. • Add reducing varnish. • Increase back-cylinder squeeze.
	Ink hangs back in the ink fountain (duct).	• Have ink manufacturer reformulate ink. • Use conical ink agitator. • Add long-flowing reducing varnish.

Troubleshooting Lithographic Ink Problems:
Insufficient Ink in Image Area

Problem	Cause	Solution
Mechanical ghosting: ghost of image from same side of sheet appears.	Ink starvation is occurring.	• Increase ink feed.
	Ink film is too thin.	• Add varnish or tint base to reduce color strength.
	Ink is too transparent.	• Add alcohol or an alcohol substitute to the dampening solution. • Decrease water feed.
	Ink rollers are the wrong diameter.	• Measure and correct rollers.
Plate blinding and plate wear: image does not take ink (gum blinding).	Gum has covered the image.	• Get another dampening solution concentrate. • Avoid excess gum and alcohol. • Adjust pH to 4.5–5.5.
	Ink does not trap on image.	• Have ink manufacturer reformulate ink.
	Paper coating deposits on image.	• Consult with papermaker. • Change papers.
	Ink has excessive drier.	• Reduce drier or change ink.
Plate blinding and plate wear: image is gone from plate.	Plate is worn out.	• Replace plate.
	Ink has dried on inkers.	• Clean press to remove this abrasive material.
	Ink is poorly ground.	• Have ink manufacturer regrind ink.
	Roller or blanket pressure is excessive.	• Check roller settings. • Check plate and blanket packing.

Troubleshooting Lithographic Ink Problems:
Insufficient Ink in Image Area

Problem	Cause	Solution
Roller stripping: ink does not properly wet the ink rollers.	Ink rollers are glazed.	• Use two-step wash treatment to clean the rollers.
	Dampening solution has excess gum or alcohol.	• Get better dampening solution concentrate.
	Ink does not wet clean rollers.	• Have ink manufacturer reformulate ink.
	Rollers carry excess dampening solution.	• Reduce dampening. • Increase ink feed.

Troubleshooting Lithographic Ink/Paper Problems

Problem	Cause	Solution
Blanket piling: ink accumulates on blanket and fails to transfer.	Ink is too tacky.	• Reduce ink tack. • Reduce the press speed. • Run more ink.
	Ink is unstable.	• Have ink manufacturer reformulate ink.
	Ink roller train cooling is insufficient.	• Lower press temperature and chill the rollers. • Get a less-volatile ink.
	Pigment is poorly dispersed or ink is poorly ground.	• Have ink manufacturer regrind ink.
	Blanket is too tacky.	• Change blankets and/or blanket wash.
	Paper coating is too weak.	• Increase water feed. • Change paper. • Add nonpiling additive. • Reduce ink tack.
	Excessive squeeze wears the paper.	• Reduce back-cylinder squeeze.

Troubleshooting Lithographic Ink/Paper Problems

Problem	Cause	Solution
Blanket piling: ink accumulates on blanket and fails to transfer *(continued)*.	Paper coating softens in water.	• Add alcohol to dampening solution. • Change paper. • Decrease water feed. • Increase the press speed.
	Paper surface has loose dust.	• Use vacuum web/sheet cleaner. • Use vacant first unit to remove dust. • Change paper.
	Ink is poorly formulated.	• Change ink.
Curl, hook, waffling: printed sheet is permanently distorted.	Ink is too tacky.	• Reduce ink tack. • Reduce the press speed. • Print heavier ink film.
	Paper is too weak.	• Replace the paper with a heavier stock.
	Blanket is too tacky.	• Use blanket with rougher surface. • Change blanket wash.
	Heavy solids appear at trailing edge of sheet.	• Turn plates around; print solids at front.
	All causes	• Use decurler on press.
Dusting: blanket lifts filler from uncoated stock or loose pigment from coated stock.	Ink is too tacky.	• Reduce ink tack. • Reduce the press speed. • Run more ink.
	Ink roller train is too hot.	• Lower temperature of chill rolls. • Order a more-stable ink. • Reduce the press speed.
	Excessive squeeze wears the paper.	• Reduce back-cylinder squeeze.
	Paper surface is too weak.	• Replace the paper. • Use vacuum web/sheet cleaner. • Use vacant first unit to remove loose lint.

Troubleshooting Lithographic Ink/Paper Problems

Problem	Cause	Solution
Dusting: blanket lifts filler from uncoated stock or loose pigment from coated stock *(continued)*.	Blanket is too tacky.	• Change blanket. • Change blanket wash. • Increase water. • Simonize or lacquer the blanket.
Gloss variation and mottle: unwanted, random variations in gloss/color appear.	Variations in paper absorbency affect holdout of ink or varnish.	• Increase ink body to improve holdout. • Decrease ink body to prevent holdout. • Reduce water feed, add alcohol. • Replace the paper. • Increase ink feed. • Apply press size/varnish to the sheet. • Print darker colors first.
	Blanket does not contact paper.	• Increase back-cylinder pressure.
Hickeys	Dirt and debris leave marks in print.	• See discussion of hickeys earlier in chapter.
Linting: blanket lifts lint from surface of paper (especially newsprint).	Ink is too tacky.	• Reduce ink tack. • Reduce the press speed. • Run more ink.
	Ink roller train is too hot.	• Lower temperature of chill rolls. • Order a more-stable ink. • Reduce the press speed.
	Excessive squeeze wears the paper.	• Reduce back-cylinder squeeze.
	Paper surface is too weak.	• Replace the paper. • Use vacuum web/sheet cleaner. • Use vacant first unit to remove loose lint.
	Blanket is too tacky.	• Change blanket. • Change blanket wash. • Increase water. • Simonize or lacquer the blanket.

Troubleshooting Lithographic Ink/Paper Problems

Problem	Cause	Solution
Picking/tearing: inked blanket pulls paper out of the surface.	Ink is too tacky.	• Reduce ink tack. • Reduce the press speed. • Run more ink.
	Blanket is too tacky.	• Change blanket and/or blanket wash. • Reduce back-cylinder pressure. • Increase water feed.
	Paper surface is too weak.	• Change paper. • Reduce ink tack. • Reduce the press speed.
Show-through: dried, printed ink produces unacceptable color on reverse side.	Paper is too transparent.	• Change to a more opaque stock.
	Ink penetrates paper.	• See the following problem: "Strike-through."
Strike-through: ink penetrates paper, producing unacceptable color on reverse side.	Ink is too soft.	• Replace the ink. • Add body varnish. • Reduce ink feed.
	Paper is too absorbent.	• Replace the paper. • Varnish the sheet.
	Paper is too thin.	• Change to heavier stock.

Troubleshooting Lithographic Ink Problems: Color Variation

Problem	Cause	Solution
Mechanical ghosting: ghost of image from same side of sheet appears.	See the problem "Mechanical ghosting" in the table "Troubleshooting Lithographic Ink Problems: Insufficient Ink in Image Area."	
Mottle: unwanted, random variations in gloss or color appear.	See the problem "Gloss variations and mottle" in the table "Troubleshooting Lithographic Ink/Paper Problems."	
Variations in ink feed: variation in color results from variations in ink or ink feed.	Water feed varies.	• Reduce water feed. • Add alcohol to dampener.
	Ink body varies.	• Be sure all additives are thoroughly mixed before placing ink on press.
	Ink color varies.	• Be sure ink is thoroughly mixed before going to press. • Check ink color on stock before going to press. • Confer with ink manufacturer.
	Ink hangs back in fountain. Yield value of ink is too high.	• Obtain a better ink. • Install a conical ink agitator. • Stir ink frequently by hand.
Variations in trap (dry): dried ink film does not properly retain wet ink from press.	Ink crystallizes.	• Decrease spray powder. • Decrease time before overprinting. • Ask ink manufacturer for trapping compound. • Ask ink manufacturer to supply ink that will not crystallize. • Avoid altering inks in pressroom.
	Wax in dried ink film is excessive.	• Avoid hard wax in first-down ink. • Work with ink manufacturer to make sure the inks meet printer's needs.

Troubleshooting Lithographic Ink Problems: Color Variation

Problem	Cause	Solution
Variations in trap (wet): second-down ink is not properly retained on printed sheet.	Ink film varies in thickness.	• Measure and control ink film thickness on press. • Run light form before heavy form. • Change color sequence.
	Ink tack varies.	• Measure ink tack. • Work with ink manufacturer to control tack.

13   Gravure Inks

With the exception of intaglio, which requires a much stiffer ink and is used for printing currency and postage stamps, most gravure printing is done by rotogravure, which is a fast, high-volume method of printing. Rotogravure presses often have a wide web and are used to print long-run magazines and catalogs, labels and wrappers, beverage carriers, wallpaper, gift wrap, textiles, linoleum, and many other materials. Sheetfed gravure is used only for a few artworks.

Rotogravure uses a cylinder that has been etched or engraved with a recessed image. The tiny cells etched or engraved into the gravure cylinder require a very fluid ink. The ink must flow readily into the cells, and the doctor blade must be able to wipe the ink cleanly from the surface of the cylinder. The ink must then flow smoothly and evenly from the cells onto the paper or other substrate.

The time and cost of preparing rotogravure cylinders are higher than the time and cost of preparing offset or flexographic plates. Because of this and other factors, short-run products are usually printed by flexography or lithography.

The rotogravure cylinder is rigid and does not conform to the substrate. If the substrate has small pits or rough areas that are larger in diameter than the individual cells, the gravure cylinder will not print on them. This problem is known as **snowflaking,** or dot skip, and is greatly reduced by using an **electrostatic assist** on press. This process facilitates contact between wet ink in the cells and the paper. In order to produce a uniform print on highly irregular or curved surfaces, offset gravure must be used.

Gravure inks must also be nonabrasive and free from grit. Abrasive particles can wear the doctor blade and the cylinder. A nicked or badly worn doctor blade will cause streaks, and the printer will have to stop the press and change blades. If the cylinder is damaged, the press may be down for a long time while the cylinder is being repaired or replaced.

Gravure inks contain volatile or low-boiling solvents that are easily dried by evaporation. Although the construction of the rotogravure press limits the exposure of ink to air, highly volatile gravure inks do evaporate on the press, and this changes their viscosity. Because of this, the printer must control ink composition (and printability and color) by checking the viscosity of the ink periodically. Modern presses are equipped with automatic viscosity controls.

Since rotogravure inks are not exposed to rubber rollers or plates, they can be made with strong solvents that help them adhere to plastic well. The solvents can be so strong that the ink actually etches the plastic.

Gravure inks are less heavily pigmented than lithographic, flexographic, or letterpress inks because the gravure cylinder delivers an ink film that is thicker than the one delivered by lithographic, flexographic, or letterpress plates. This thicker ink film delivered in shadow areas contributes to the gloss that is characteristic of "gravure quality." Gravure inks are not shipped at the low viscosities that are used on press because the pigments would tend to settle out. Instead, gravure inks are usually sold in concentrated form to be diluted, or "let down," at press side, often utilizing recovered solvent in the process.

Because the gravure cell transfers a large volume of ink with a low pigment concentration, gravure is an excellent method for printing gold (bronze) and silver (aluminum) inks. Gravure golds are brilliant and nontarnishing. These metallic pigments are difficult to print by lithography because of the high pigment loadings required for lithographic printing.

Since gravure inks for package printing and publication printing differ in composition and have different end-use requirements, they are often considered separately. Plastic film makes an excellent packaging substrate for rotogravure because it is smooth, glossy, and nonabsorbent. With transparent plastic film, the image is usually printed on the reverse side, which gives the print a high gloss and protects the ink film.

Solvency and Viscosity

The behavior of a solvent depends on the properties of the other materials in a mixture. In general, a given solvent will only dissolve a particular class of binders or vehicles.

Solvency is also related to viscosity: for a given amount of solvent, the stronger it is, the lower the viscosity of the resulting solution. For example, the viscosity of an 8% solution of nitrocellulose in ethyl acetate is lower than the viscosity of an 8% solution of nitrocellulose in normal butyl acetate. Ethyl acetate is a much better solvent for nitrocellulose than is normal butyl acetate.

The typical solvents used for gravure inks and the properties of some of those solvents are listed in the tables on pages 184 and 185.

Classification of Gravure Inks

Gravure inks are commonly classified according to their application and composition. Type A and B inks are used for publication printing, and types C through X for packaging. Any of these inks may be formulated with wax or synthetic resin to achieve scuff resistance or with other additives to achieve other properties.

Representative gravure ink formulations are presented in the accompanying table.

Typical formulations of rotogravure inks as shipped by the manufacturer

General	Publication	Packaging
Parts	Parts	Parts
20 Pigment and toner	25 Gilsonite	33 Titanium dioxide
28 Resin	2 Carbon black	22 Nitrocellulose varnish (in ethanol/toluol/ester)
47 Solvent	2 Toner	
5 Wax, plasticizer, and additives	6 Metal resinate	
	3 Wax compound	37 Rosin ester varnish
100	46 Lactol spirit	
	16 Toluol	3 Plasticizer
	100	5 Wax compound
		100

Gravure Inks for Various Substrates

In addition to these publication and packaging inks, gravure is sometimes used with heat-transfer inks for textiles. Heat transfer inks contain a sublimable dye. The dye is printed on paper using gravure, lithography, flexography, or screen printing. When the printed paper is heated and pressed against a textile, the dye sublimes (evaporates) and migrates into the textile fibers, decorating the textile.

As with other types of printing, the printer must consult the ink manufacturer for help in selecting the best ink. Newspaper supplements are usually printed with type A inks. Supercalendered or coated publication papers are usually printed with type B inks.

For glassine and vegetable parchment papers, a good film former is needed for good flexibility and adhesion, and type E inks work very well; the lack of residual odor is an additional advantage if the paper is used for a food or candy wrap.

Typical solvents for gravure ink

Chart by Leon Knorps of Flint Ink Corp. and reprinted from the Gravure Association of America Solvent Manual

Type	Uses	Fast	Normal	Slow	Binder
Publication					
Group VII (or Type A)	Uncoated-paper inks for newspapers and other publications	Lactol spirits or textile spirits	Blends of lactol spirits and toluene, or VM&P and toluene	VM&P	Metal resinates and hydrocarbon resins
Group VI (or Type B)	Coated-paper inks for catalogs, magazines, and other publications	Textile spirits	Blends of lactol spirits and toluene, or VM&P and toluene	VM&P	Ethyl cellulose, metal resinates, phenolics, and hydrocarbon resins
P	Proofing ink for coated and uncoated paper	Lactol spirits	Blends of VM&P, toluene, and xylene	Xylene	Same as production inks
W	Water ink for coated and uncoated catalogs, newspapers, magazines, and other publications	Water	Water	Water	Various resin emulsions
Packaging					
C	Inks for paper and nitrocellulose-coated cellophane, metallized paper, and aluminum foil. End uses are folding cartons, labels, and flexible packaging	Ethyl acetate, methyl ketone, or acetone	Isopropyl acetate (ethyl alcohol or isopropyl alcohol may be used with ethyl or isopropyl acetate)	N-propyl acetate, N-butyl acetate	Nitrocellulose plus modifying resins
D	Tough top lacquers for paper and paperboard; inks for polymer-coated cellophane, polyethylene, and polyester	Ethyl alcohol (may require some toluene if a cosolvent resin is used); ethyl acetate	Isopropyl alcohol (may require some toluene if a cosolvent resin is used), isopropyl acetate	N-propyl alcohol (may require some toluene if a cosolvent resin is used), N-propyl acetate	Polyamide (may be modified with nitrocellulose)
E	Low-odor inks and lacquers for paper and paperboard; inks for nitrocellulose-coated cellophane and some aluminum foils	Methyl alcohol or ethyl acetate	Ethyl alcohol, isopropyl alcohol, or isopropyl acetate	N-propyl alcohol or N-propyl acetate	Alcohol-soluble nitrocellulose plus modifying resins
M	Low-cost lacquers for cartons and labels	Blends of toluene, hexane, and methyl ethyl ketone	Toluene	Xylene	Polystyrene
T	Inks for paper and paperboard, especially high-gloss inks with no top lacquer; alcohol-resistant inks	Methyl ethyl ketone, ethyl acetate	Toluene, isopropyl acetate, N-propyl acetate	Xylene	Chlorinated rubber
V	Inks for vinyl substrates such as wallcoverings and vinyl films to be laminated to wallboard	Methyl ethyl ketone, acetone	Methyl ethyl ketone plus methyl isobutyl ketone	Methyl isobutyl ketone	Polyvinyl chloride, copolymer resins
W	Water inks for all purposes	Water and ethyl alcohol	Water	Water	Various solution resins and/or emulsions
X	Inks that do not fall in other categories; fluorescent inks, barrier coatings, and certain film inks		No generalizations possible		

Physical constants of gravure solvents	Evaporation Rate*	Boiling Range (°F)	Flash Point† (°F)	Pound per Gallon @ 68°F
Esters				
Methyl acetate	5	126–136	14	7.55
Ethyl acetate	10	162–176	24	7.40
Isopropyl acetate	12	183–194	40	7.25
Normal propyl acetate	22	203–217	58	7.35
Secondary butyl acetate	33	219–243	72	7.18
Isobutyl acetate	36	230–246	64	7.20
Normal butyl acetate	63	239–266	72	7.30
Amyl acetate	100	248–302	77	7.20
Cellosolve acetate	330	293–329	124	8.10
Alcohols				
Methyl alcohol	10	147–149	52	6.60
Solvent alcohol (ethyl)	20	167–176	63	6.80
Isopropyl alcohol	27	178–181	53	6.55
Normal propyl alcohol	55	203–208	77	6.70
Secondary butyl alcohol	63	210–212	75	6.68
Isobutyl alcohol	77	223–228	82	6.68
Normal butyl alcohol	125	241–246	84	6.75
Glycol ethers				
Dowanol PM	88	245–259	100 (oc)	7.65
Methyl Cellosolve	130	253–259	107	8.03
Cellosolve	190	270–279	104	7.74
Butyl Cellosolve	1,000	331–343	160	7.51
Ketones				
Acetone	5	133–135	0	6.60
Methyl ethyl ketone	11	172–178	21	6.71
Methyl isobutyl ketone	37	237–243	73	6.68
Methyl butyl ketone	50	235–266	83 (oc)	6.75
Cyclohexanone	270	266–343	111 (oc)	7.88
Aromatic hydrocarbons				
Toluol	26	229–233	40	7.24
Xylol	88	275–290	81	7.26
Aliphatic hydrocarbons				
Hexane	7	150–158	–7	5.70
Fast-diluent naphtha	7	140–180	32‡	5.75
Heptane	15	200–220	25	6.05
Lacquer diluent	17	200–240	30‡	6.20
Octane	31	215–230	56	6.20
VM&P naphtha	51	215–300	52	6.30
Mineral spirits	600	310–400	104	6.50
Nitroparaffin				
2-nitropropane	50	246–248	103 (oc)	7.24

*The evaporation rate is a comparison based on ethyl acetate, which has a standard rating of 10. Isopropyl acetate, rated as 12, evaporates slightly slower; acetone, rated at 5, evaporates faster than ethyl acetate.

†Flash points are closed cup as reported in the *Fire Protection Guide on Hazardous Materials* (published by the National Fire Protection Association), except where "(oc)" follows the flash point, in which case the flash point reading indicates open cup. Open cup readings are approximately 10°F higher than closed cup readings.

‡At very low temperatures, flash point determinations are inconsistent. The double-dagger denotes that flash occurs below the temperature shown.

For printing poly(vinyl chloride), type V inks are preferred; the strong solvents permit use of resins required for good adhesion. For poly(vinylidene chloride) (sometimes called PVDC) and for "K film" (cellophane coated with PVDC), type D inks are often used.

Aluminum foil is printed with type C or D inks, but to ensure good adhesion, the printer must buy a foil that has been degreased. The foil may be primed with nitrocellulose, shellac, or a vinyl resin.

Testing Gravure Inks

The development and application of a significant testing procedure is not a simple task. However, a simple drawdown with a broad scraper or knife will give a fair amount of information about an ink. The drawdown can conveniently be done with a wire-wound rod (a Meyer bar) or, better, a "K" proof. The sample ink can be drawn down next to a standard, and the inks compared for drying speed, shade, strength, finish, and penetration.

A more sophisticated instrument is the GRI printability tester, which is a bench-size proof press. It has been used in product development and research laboratories to study gravure ink/paper interactions.

When ordering ink, the printer may wish to specify uniformity of strength and shade, shipping viscosity, density or specific gravity (weight per gallon), and dilution viscosity. Viscosity, of course, is monitored not only on dilution but also throughout the printing process.

Appropriate quality control tests for rotogravure inks

Color	Masstone and Undertone	Skid Resistance
Viscosity	Bleeding into Plastic	Product Resistance
Lightfastness	Tinctorial Strength	Flexibility
Flash Point	Opacity/Hiding Power	Density and
Gloss	Drying	Specific Gravity
Odor	Rub and Scuff Resistance	Adhesion
Fineness of Grind		

Troubleshooting Gravure Ink Problems: Publication Gravure

Problem	Cause	Solution
Cylinder wear.	Mechanical problems may include excessive blade pressure or lack of blade oscillation.	• Put cylinder and press into top shape.
	Ink contains foreign particles.	• Filter ink.
	Ink is drying too rapidly.	• Add slow solvent.
	Pigment is abrasive.	• Have ink manufacturer reformulate ink.
	Pigment dispersion is poor.	• Have ink manufacturer regrind ink.
	Paper stock is abrasive.	• Replace the paper. • Use sheet cleaner.
Fuzziness of print.	Static electricity forms on the web.	• Use static eliminator. • Moisten the web. • Ask ink manufacturer to supply a more polar solvent. • Increase ink viscosity.
Impression roll buildup: ink coats on impression roll.	Impression roll durometer is too high when printing on coated paper.	• Use impression roll with lower durometer. • Add wax compound to ink.
Mottle or crawl; ink spreads on print, creating undesirable pattern.	Ink dries too slowly.	• Add a faster solvent.
	Ink is too diluted.	• Add fresh ink and/or extender.
	Press speed is too slow.	• Get press up to speed. • Reduce impression pressure. • Use thinner doctor blade. • Adjust blade to wipe at a sharper angle.
Pinholes in print (occurs when printing tissue).	Penetration is excessive.	• Change paper. • Have ink manufacturer reformulate ink.

Troubleshooting Gravure Ink Problems: Publication Gravure

Problem	Cause	Solution
Poor adhesion: dried ink comes off.	Binding is insufficient.	• Have ink manufacturer reformulate ink. • Add binding varnish.
	Ink film flexibility is poor.	• Have ink manufacturer reformulate ink. • Decrease ink pigmentation.
Poor trapping: ink film applied over a printed film is not retained.	First-down or underlying ink is not dry.	• Have ink manufacturer reformulate ink. • Adjust drying temperature. • Reduce viscosity of overprint inks.
	First-down or underlying ink is too glossy.	• Have ink manufacturer reformulate ink. • Increase pigmentation of first-down ink.
Screening: a screen pattern shows up in the dried print.	Ink is too viscous.	• Reduce ink viscosity.
	Ink dries too fast.	• Add a slower solvent.
	Ink body is too short.	• Add varnish extender.
	Impression is insufficient.	• Increase impression. • Ease doctor blade pressure. • Control temperature of paper and fountain ink.
Slow drying, blocking, or picking.	Rewind is too hot.	• Reduce drying temperature.
	Rewind is too tight.	• Adjust rewind tension.
	Drying is too slow.	• Ask ink manufacturer for ink that is faster and block-resistant. • Increase drying temperature.

Troubleshooting Gravure Ink Problems: Publication Gravure

Problem	Cause	Solution
Slow drying, blocking, or picking *(continued)*.	Resin has poor solvent release.	• Add faster solvent. • Have ink manufacturer reformulate ink.
	Ink contains too much plasticizer.	• Have ink manufacturer reformulate ink.
Speckle, skipped dots, or missing halftone dots on the print.	Paper is uneven.	• Use a smoother paper. • Install electroassist. • Increase impression pressure. • Soften paper with steam.
	Ink is too viscous.	• Reduce ink viscosity. • Add varnish extender.
Streaks or railroads: ink does not wipe cleanly from cylinder.	Ink is too viscous.	• Reduce viscosity.
	Ink dries too fast.	• Add slower solvent.
	Ink is too heavily pigmented.	• Add varnish.
	Ink contains foreign particles.	• Filter ink. • Contact ink manufacturer.
	Doctor blade and oscillation are improperly adjusted.	• Check oscillation of doctor blade. • Change doctor blade angle. • Polish blade and cylinder. • Change blade pressure.
Surface smear: ink is easily smeared.	Drying is slow.	• Have ink manufacturer reformulate ink. • Adjust drying temperature.
	Solvent release is poor.	• Have ink manufacturer reformulate ink.

Troubleshooting Gravure Ink Problems: Packaging Gravure

Problem	Cause	Solution
Abrasion: cylinder wear.	Pigment is unground or abrasive.	• Ask ink manufacturer to regrind or reformulate the ink.
	Solvent is too fast.	• Add slower solvent.
	Steel or porcelain is present in ink.	• Filter or remake ink.
	Cylinder and doctor blade are not adjusted properly.	• Adjust, remake, or rechrome the cylinder. • Adjust the doctor blade.
Adhesion: ink and/or varnish do not bind to each other or to substrate.	Film or foil not treated adequately.	• Check film or foil for treatment. • Consult supplier of film or foil.
	Ink does not contain sufficient plasticizer.	• Consult ink supplier.
	Ink contains excessive plasticizer.	• Have ink manufacturer reformulate ink.
Bleed: one color bleeds into another.	Ink is too slow.	• Use faster solvent.
	Drying temperature is too low.	• Raise the temperature.
	More air is required.	• Clock air velocity.
	Ink was improperly formulated.	• Have ink manufacturer reformulate ink.
Blocking: undesired adhesion between surfaces.	Drying is not sufficient.	• Raise the temperature.
	Solvent is trapped in the ink.	• Correct solvent balance. • Use faster solvent.
	Web rewound too warm.	• Use chill rollers.
	Rewind has excessive pressure.	• Reduce tension.

Troubleshooting Gravure Ink Problems: Packaging Gravure

Problem	Cause	Solution
Blocking: undesired adhesion between surfaces *(continued)*.	Ink or varnish is over-plasticized.	• Have ink manufacturer reformulate ink.
	Binders have low melting points.	• Have ink manufacturer reformulate ink.
Brittleness: substrate breaks when flexed.	Drying system is excessively hot.	• Control web temperature.
	Substrate lost moisture.	• Introduce moisture.
	Plasticizer migrated.	• Reduce temperature. • Add faster solvent. • Check film supplier.
Cobwebs: filmy, weblike buildup on doctor blade, impression roll, engraving, or press frame.	Air drafts at nip dry ink prematurely.	• Stop drafts.
	Ink dries too fast.	• Add slower solvent.
	Viscosity is high.	• Reduce viscosity to normal level.
Comets and darts: ink deposited in shapes of comets and darts.	Cylinder contains a defect.	• Repair and/or polish cylinder. • Check engraver's proofs.
	A foreign substance is under doctor blade.	• Strain ink.
	Ink is too dry.	• Add lubricant to ink.
Crawling or mottle: poor lay of ink.	Ink does not wet substrate evenly.	• Change blade angle. • Change solvents.
	Viscosity is low.	• Increase viscosity.
	A poor pigment was selected.	• Have ink manufacturer reformulate ink.
	Press is too slow.	• Increase speed.
	Cylinder is too deep.	• Reetch cylinder.
	Pressure is too high.	• Adjust pressure.

Troubleshooting Gravure Ink Problems: Packaging Gravure

Problem	Cause	Solution
Doughnut: center missing from screen dot.	Drying is too fast.	• Add slow solvents.
Drag-out, or slur: bead of ink appears at trailing edge of print.	Ink viscosity is low.	• Increase viscosity.
	Blade angle is wrong.	• Align blade properly.
	Doctor blade is wavy.	• Adjust and clean blade. • Adjust backup blade.
	Tension control is poor.	• Check tension.
	Ink dries too slowly.	• Add faster solvent. • Increase drying temperature.
Drying-in, or weak print.	Ink viscosity is too high.	• Reduce viscosity.
	Ink is short-bodied.	• Add clear extender.
	Air drafts at nip.	• Stop drafts.
	Ink is drying too fast.	• Add slower solvent. • Reduce dryer temperature.
	Wiping is excessive.	• Adjust doctor blade.
	Pressure is excessive.	• Reduce pressure.
	Press speed is too fast.	• Reduce press speed.
Foaming: small bubbles in ink.	Surface tension is excessive.	• Have ink manufacturer reformulate ink. • Use antifoaming agent.
	Ink pump is operating too vigorously.	• Reduce ink flow. • Reduce free-fall of ink.
Grainy print: print not smooth.	Stock is not smooth.	• Replace stock. • Use electroassist.
	Ink has high viscosity.	• Reduce viscosity.
	Press is too slow.	• Increase speed.

Troubleshooting Flexographic Ink Problems: Packaging Gravure

Problem	Cause	Solution
Grainy print: print not smooth *(continued)*.	Stock is too dry.	• Introduce moisture. • Reduce temperature of stock. • Check cylinder pressure.
Hazy film: slight opacity in clear, unprinted area of the film.	Cylinder has slight roughness.	• Polish cylinder.
	Wipe is poor.	• Check doctor blade setup.
Hazy ink: foggy appearance in ink film.	Ink is overpigmented.	• Add clear extender.
	Humidity is too high.	• Slow down ink.
	Chrome job is poor.	• Remake cylinder.
	Solvent combination is improper.	• Change solvents. • Consult ink manufacturer.
	Tension is improper.	• Check tension on stock.
Offsetting: transfer of the printed matter to the reverse side of sheet or web.	Ink does not dry soon enough.	• Check dryer and air exhaust.
	Ink film is wet or tacky.	• Add faster solvent. • Increase dryer temperature.
Picking: lifting of spots of ink from printed area.	Ink is slow-drying.	• Add faster solvent.
	Heat is insufficient.	• Increase heat and air velocity.
	Ink is too viscous.	• Reduce viscosity.
Picking in multicolor work: previous ink picks off sheet or on roller.	First-down ink is too slow.	• Add faster solvent.
	Second-down ink has decreased viscosity.	• Use extender.
	Doctor blade needs adjustment.	• Adjust doctor blade. • Use air blast on offending cylinder.

Troubleshooting Gravure Ink Problems: Packaging Gravure

Problem	Cause	Solution
Pinholes: appearance of small holes in the printed area.	Ink has imperfections.	• Reduce viscosity. • Add a more-active solvent. • Have ink manufacturer reformulate ink. • Adjust blade angle.
	Stock has imperfections.	• Use electroassist. • Get better stock.
Railroads (comets): continuous line showing in the unprinted areas.	Particles are lodged under doctor blade.	• Filter ink.
	Scratches are on cylinder.	• Polish or remake cylinder.
	Scratches are on blade.	• Replace doctor blade.
	Cylinder has a high spot or chrome deposits.	• Remove burr and polish cylinder.
Screening: screen pattern in dried print.	Ink viscosity is too high.	• Lower viscosity.
	Ink dries too fast.	• Add slower solvent.
	Blade angle is too sharp.	• Flatten blade angle.
Skips: engraving dots that have not printed.	Printing surface is rough.	• Use primer coats.
	Pressure is insufficient.	• Check pressure.
	Ink dries too soon: "drying-in."	• Add slower solvent.
	Ink is not on cylinder.	• Check applicator.
	Ink viscosity is too high.	• Add solvent.
	Ink has poor flow.	• Check circulation pump.

Troubleshooting Gravure Ink Problems: Packaging Gravure

Problem	Cause	Solution
Snowflakes: random, minute, unprinted areas.	Stock is rough.	• Select a better stock. • Use electroassist. • Add polar solvent.
	Ink viscosity is too high.	• Add solvent.
	Ink dries too fast.	• Add slow solvent.
Static: fuzzy print, fuzz hairs.	Web has static.	• Use static eliminator. • Use tinsel. • Increase humidity. • Add steam vapor.
	Ink viscosity is too low.	• Increase viscosity. • Use pigmented extender. • Add polar solvents.
	Electroassist is excessive.	• Turn down or turn off electroassist.
Streaking appearing at one spot.	Cylinder is damaged.	• Polish or remake cylinder.
Streaking coming and going.	Foreign particles and lint are present.	• Wipe doctor blade. • Filter ink.
Streaking following oscillation.	Blade is damaged.	• Polish doctor blade.
Streaking: small streaks at trailing edge of print.	Ink viscosity is too low.	• Add varnish.
	Cells are overetched.	• Have cylinder checked.
	Static electricity is present.	• Use static eliminator. • Increase humidity.
Volcano: broken surface in print resembling volcano.	Solvent trap is faulty.	• Add faster solvent to first-down ink. • Have ink manufacturer reformulate ink. • Change solvents.
	Dryer is too hot.	• Reduce heat.

Troubleshooting Gravure Ink Problems: Packaging Gravure

Problem	Cause	Solution
Wiping: dirty print from uncleaned cylinder.	Doctor blade performs improperly.	• Check blade setting and condition.
	Ink is highly viscous.	• Add thinner.
	Ink dries too fast.	• Add slower solvent.
	Ink is highly pigmented.	• Add clear extender. • Have ink manufacturer reformulate ink.
	Ink viscosity is too low.	• Add varnish. • Increase blade pressure.

14 Flexographic Inks

Flexography is a method of direct rotary printing from resilient relief plates (rubber, synthetic rubber, plastic, photopolymer, and synthetic polymer) that carry fluid inks to virtually any substrate. It is named for the "flexible" plates that are used to carry the image. The process is often described in conjunction with letterpress, which is misleading because, except for the fact that the image area is a raised surface, flexography bears little similarity to letterpress. Letterpress inks are stiff, pasty, and difficult to dry. These inks are compatible with the complicated letterpress distribution system. The simpler flexographic inking system uses low-viscosity fluid inks that dry easily because of their high volatility.

Prior to 1952, flexography was called "aniline printing" because coal tar dyes, derived from aniline oil, were originally used as the chief colorants. By 1950, pigments were more commonly used than the aniline dyes, and a new name was needed. The difference between dyes and pigments is that dyes dissolve in the vehicle or solvent while pigments are insoluble. Even though pigments resist fading and bleeding better than dyes do, dyestuffs are still used when maximum transparency and brilliance are required.

Flexographic printing technology is advancing rapidly in the 1990s. Under the best commercial conditions, flexography produces print quality comparable to lithography or gravure. It is used to print packaging materials such as beverage carriers and milk cartons, labels, corrugated boxes, and flexible packaging (film and foils). It is also used to print paperback books. In the late 1980s, flexography was being used increasingly to print newspapers, newspaper inserts (freestanding inserts), Sunday comics, business forms, and other products printed on newsprint. The bulk of flexographic printing is applied to wide web (e.g., polyethylene wrappers, polyester, and polyolefin films like polypropylene, others like nylon and cellophane, and corrugated linerboard), narrow web (e.g., pressure-sensitive labels), and corrugated boxes. This wide variety of substrates requires a broad range of ink formulations. However, attempts to classify flexo inks have not been successful.

Viscosity   Like gravure inks, flexo inks have a low viscosity, and if they were shipped and stored at the viscosity used for

Typical
formulations of
flexographic inks

White for Flexible Packaging	**Red for Paper**
Parts	Parts
33 Titanium dioxide	24 Red lake C chip
10 Nitrocellulose varnish	35 Ethanol
2 Ethyl acetate	12 Isopropyl acetate
2 Plasticizer	15 Maleic resin varnish
3 Cellosolve	9 Nitrocellulose varnish
20 Maleated rosin ester	2 Wax compound
30 Ethanol	3 Plasticizer
———	———
100	100

printing, the pigments would rapidly settle out and printers would not be able to adjust the ink strength. The inks, therefore, are shipped in a concentrated form and "let down" to running viscosity at press side, using a solvent specified by the ink manufacturer. The viscosity reflects both pigment concentration and ink flow. Close control of viscosity is essential to good color control. Flexo inks, like gravure inks, evaporate on press. Evaporation changes the viscosity and color value of the ink. The printer controls the flow and color of the ink by monitoring the viscosity and adding solvent as required. Continuous monitoring and automatic solvent addition are often more beneficial economically than manual control, which is erratic.

The advent of the reverse-angle doctor blade has greatly increased the viscosity of the ink that the flexo press can handle, thereby enhancing the versatility of the process. Using the reverse-angle doctor blade also increases the control of the ink film applied to the plate.

Viscosity Control

Variation in viscosity and color strength is the major cause of color variation in flexographic printing. Flexo inks must have a low viscosity to transfer properly in the inking system. However, if the viscosity is too low, the ink will not remain on the surface of the plate or have adequate density. It may run down the sides of the image, fill in reverses and halftones, or even build a halo around the printed letter or image.

An ink formulated with the correct balance of solvents prints at a lower viscosity than an improperly formulated ink. As solvent evaporates from the ink fountain and rollers, the solvent balance changes because the more volatile solvents evaporate more readily. The printer should consult the ink manufacturer for the proper solvent mixture to replace evaporating solvent.

There is an additional viscosity concern with aqueous inks which contain a binder that is made soluble by a base, such as ammonia or morpholine. If insufficient base is present, the binder will begin to precipitate or separate from the solution before the ink is printed. When this happens, the pH decreases and the viscosity of the solution increases causing the ink to coagulate. Amines are volatile just like solvents and will evaporate while the ink is on the press. Evaporation of amines will also cause the pH of the ink to decrease and result in coagulation. The pH should be monitored while running the ink so that additional base or pH conditioner can be added at press side if needed.

Although viscosity is usually measured with efflux cups, several automatic viscosity control systems are available that automatically maintain the optimum viscosity of inks.

Drying Problems

Flexo inks for printing on film are usually dried by evaporation of the solvent. Inks for paper and boxboard are often water-based. They dry, at least partly, by absorption of the water by the paper. The pigment and binder are left on the surface and the remaining solvents are dried by evaporation that may be heat induced. As with any other ink, the drying temperature of a flexo ink must be controlled. If the temperature is too high, the ink film may soften. If the temperature is too low, the solvent may be incompletely removed. Both extremes may cause the soft ink to block in the rolls or between the sheets, or smear in the finishing processes. Hot ink films can be cooled on a chill roll, but it is better to keep the process under control—to supply just enough heat to the web to dry the ink.

Drying of flexographic prints can also be controlled by adding either a "faster" solvent to the ink to accelerate drying or a "slow" solvent to retard drying. Solvents may be classed as "slow," "normal," and "fast" as shown in the accompanying table.

The relative speeds of various solvents

Slow	Normal	Fast
n-Propyl alcohol	Ethyl alcohol	Isopropyl acetate
n-Propyl acetate	Isopropyl alcohol	Ethyl acetate
Cellosolve		

In multicolor printing, several colors are printed in sequence, building ink layers on the print and influencing drying. The drying rates of the inks must be graduated by selecting the right solvents so that the inks will trap properly. The first-down ink usually contains the fastest solvent, the last-down, the slowest.

Solvents used in flexography

Solvent	Weight* Density	Flash Point (closed cup) °F	°C	Evaporation Rate†
High-flash naphtha	6.56	100	38	0.1
Isopropyl alcohol	6.58	53	12	2.3
Ethyl alcohol	6.57	55	13	3.3
N-propyl alcohol	6.66	81	27	1.1
Ethyl acetate	7.49	24	−4	6.2
Isopropyl acetate	7.24	40	4	5.0
N-propyl acetate	7.40	58	14	2.8
Ethylene glycol ethyl ether (Cellosolve)	7.74	110	43	0.4
Water	8.33	——	——	——

Source: *NPIRI Handbook*
*Pounds per gallon at 68°F (20°C)
†Butyl acetate = 1.0

Water-Based Inks

Water-based inks were first formulated from styrene-maleic or rosin-maleic resins and a base such as ammonia or morpholine. These inks did not provide any gloss and had poor rub resistance. Most water-based flexo inks are now made from acrylic resins, either in solution or in emulsion (suspension) forms. They have good gloss and rub resistance, but the solutions tend to be too viscous, and the emulsions tend to be unstable. Water-based inks for nonabsorbent substrates like polyethylene, polypropylene, and polyester films dry by evaporation since the substrates are not porous and can not absorb the inks. Water evaporates very slowly, making the inks hard to dry. However, with good raw materials and careful formulation, useful inks are produced.

Water-based flexo inks are used for a variety of end uses and must be formulated for the requirements of the job.

News inks. The outstanding properties of newspapers printed by flexo are due mostly to the attributes of the ink. Flexo news inks are based on acrylic polymers with free acid groups that are brought into solution with ammonia or an amine. When the ink is printed, alum in the newsprint precipitates the acrylic binder, giving excellent rub resistance and holdout that results in bright colors and high opacity. These news inks have a higher rub resistance than any other news ink, and, because the ink is held out on the surface, less show-through is seen. This minimal amount of show-through permits use of a sheet that is lower in weight than required for printing by offset or letterpress.

Water-based inks for other substrates. Since gloss is not important for paper bags and conventional brown corrugated boxes, inexpensive water-based inks perform well. On the other hand, good gloss and rub resistance are very important for coated, bleached top liner. Whether the top liner is preprinted (printed before the corrugated board is manufactured) or the box is printed after manufacture, a carefully formulated ink is required.

For food packaging, it is extremely important that an ink be odorless. Water-based inks may be suitable, but a major source of odor from water-based inks are the residual monomer odors from the resins themselves. Flexographic inks based on nitrocellulose must be modified with plasticizers and other resins that may produce odors. The ink manufacturer should be consulted if freedom from residual odor is important.

Plugging or Fill-In

Filling in of halftone dots on the plate is a serious problem that occurs with water-based flexo publication inks. Although it occurs with water-based flexo inks, it does not appear to be caused entirely by the ink, but the result of the interaction between the ink, paper, and plate. Some paper stocks perform better than others. The fill-in may be the result of many single factors or a combination of them:

- Alum in the newsprint may destabilize or precipitate the resin in the ink.
- Dirt in the ink and dried ink particles from the ink or press aggravate fill-in.
- The binder may pick up lint or fibers from the paper, filling in the halftone dots.
- Plates that soften or swell in the presence of water and solvents in the ink contribute to the problem.

Improper printing techniques such as inadequate control of ink viscosity, improper printing impression, and an incorrect doctor blade setting also contribute to fill-in.

Flexo Inks for Various Substrates

Solvent inks are often used on coated papers or lightweight stock. Polyethylene-coated paper and board, like polyethylene films, require a flame- or corona-discharge treatment if ink is to adhere to them. Alcohol-based polyamide inks are preferred. Polyolefin films, nitrocellulose, and polymer-coated nitrocellulose are printed with polyamide inks that give excellent adhesion, gloss, and flexibility. Addition of nitrocellulose increases grease and heat resistance. A polyamide cosolvent (alcohol and aliphatic hydrocarbon) ink gives good resistance to water, acid, and alkali.

Poly(vinyl chloride) films always contain some plasticizer. This may bleed back into the ink, soften it, and cause it to block. The ink manufacturer should be consulted before trying to print this (or any other) film.

Alcohol-reducible nitrocellulose and polyamide inks are used to print aluminum foil. The bright, metallic sparkle of aluminum foil is retained by printing with colors of maximum transparency. To promote good ink adhesion, the manufacturer of the aluminum foil removes any grease or oil and treats the foil with a wash coat of nitrocellulose or shellac.

Flexography prints gold and silver inks (bronze and aluminum) reasonably well. As with fluorescent inks, the thicker the ink film, the better the ink prints.

Surface Printing, Reverse Printing, and Laminating

Certain printing operations are possible when printing on a transparent film that are impossible when printing on paper or board. Surface printing can be applied to any substrate, but flexographic inks, like gravure inks, can be printed on the underside of the film, a process known as "reverse" printing.

Combinations of films or foil on paper (called laminates) provide materials with protection from oxygen, moisture, or grease. This degree of protection cannot be achieved with a single material, and it is sometimes desirable to print in between the layers. The ink is reverse-printed on the film before the other material is laminated to it. Inks used in this way must adhere well to both materials used in the laminate. The ink must have low solvent retention because solvent must not be trapped between the layers, but gloss and rub resistance are not important.

Flexographic Plates

Flexo plates were first made from natural rubber and then plastic, both of which swell readily when exposed to many solvents. Because of this, ink solvents were largely restricted to water, alcohol, and aliphatic hydrocarbons. Aromatic hydrocarbons (toluene) and ketones (MEK or MIBK) were unsuitable, and esters (ethyl acetate) had to be diluted with less active solvents. The resins that could be used as binders were also limited. The ones most commonly used were shellac, rosin esters, alcohol-soluble nitrocellulose, and styrene-maleic resins. Synthetic rubber plates, like nitrile rubber or NBR (nitrile butadiene rubber) and chloroprene or CR rubber, can be used with a wider range of solvents and resins.

Although rubber and synthetic rubber plates are still used, a great deal of flexo printing is done from photopolymer plates. Photopolymer plates resist solvents better than rubber plates do. This feature makes them more versatile because the printer and ink manufacturer have a wider range of solvents from which to choose.

Testing Flexographic Inks

As a minimum, the ink should be applied to the substrate using an anilox hand proofer. The print made in this way can be examined for color and gloss, adhesion, rub resistance, and other properties. Appropriate quality control tests are listed in the accompanying table. Opacity, color strength, and other tests are described in Chapter 10, "Testing."

Appropriate quality control tests for flexographic inks

Wet Ink Tests	**Dried Ink Film Tests**
Color	Color
Tinctorial Strength	Product Resistance
Opacity/Hiding Power	Opacity/Hiding Power
Viscosity	Scuff Resistance
Drying	Adhesion
Masstone and Undertone	Flexibility
Bleeding into Plastic	Flash Point
Fineness of Grind	Odor
Density and Specific Gravity	Heat Resistance
Wet Rub Resistance	Rub Resistance
Flash Point	Gloss
Block Resistance	Skid Resistance
pH	Fade Resistance
Surface Tension	Ice Water and Freeze Resistance

**Trouble-
shooting**

Because flexography is used for a wide variety of products, troubleshooting is especially difficult. As with all printing processes, the ink and substrate interact; problems resulting from this interaction are more conveniently solved by changing the ink than by changing the substrate. However, changes in the ink must be carefully considered so that solving one problem does not produce another. Some guides for troubleshooting flexographic inks are given in the accompanying table.

Troubleshooting Flexographic Ink Problems

Problem	Cause	Solution
Blocking: sheets stick together after drying.	Ink is not dry.	• Raise drying temperature. • Add faster solvent.
	Ink is improperly formulated; there is too much plasticizer.	• Get another ink from ink manufacturer.
	Rewind tension is excessive.	• Control rewind tension.
	Storage temperature is excessive.	• Store printed rolls in a cool place.
Color varies between repeat runs.	Anilox roller is worn.	• Get new anilox roller. • Check doctor blade setting. • Check density of printed solids and adjust viscosity. • Use ceramic roller.
	Ink color varies.	• Ask ink manufacturer to correct color.
Color varies during the run.	Evaporation of solvent changes color strength.	• Monitor viscosity more carefully and add correct amount of solvent. • Ask ink manufacturer for correct solvent.
Feathering: edges around image area are ragged.	Ink dries too fast.	• Add slower solvent. • Avoid drafts on plate or ink fountain.
	Pressure on plate is excessive.	• Adjust pressures correctly.
	Ink viscosity is uncontrolled.	• Maintain recommended viscosity.
Fill-in of reverses or halftones: shadows become solids.	Plate pressure is excessive.	• Adjust roller to recommended pressure.
	Ink is dirty.	• Filter or replace the ink.

Troubleshooting Flexographic Ink Problems

Problem	Cause	Solution
Fill-in of reverses or halftones: shadows become solids *(continued)*.	Anilox-to-plate pressure is excessive.	• Adjust roller to recommended pressure.
	Ink feed is excessive.	• Reduce ink feed.
	Ink viscosity is too high.	• Add solvent.
	Ink dries too fast.	• Add slower solvent. • Speed up the press.
	Plates are sensitive to water.	• Consult the platemaker.
	Ink is too sensitive to paper.	• Consult the ink manufacturer. • Add base to ink.
	Paper is linting or contains too much alum.	• Consult the papermaker.
	Ink is poorly ground.	• Return ink to ink manufacturer. • Regrind or filter ink.
	Plate is too shallow.	• Get better plate. • Reengrave plate.
Foaming.	Ink falls over open space.	• Close up system: deliver falling ink under surface of pool. • Force foamy ink through filter. • Add defoamer. • Consult ink manufacturer.
Halos: outline forms around type or image.	Pressure is excessive.	• Adjust pressures.
	Ink feed is excessive.	• Reduce ink feed.

Troubleshooting Flexographic Ink Problems

Problem	Cause	Solution
Ink adheres poorly to substrate.	Surface treatment is poor, especially on polyolefin or aluminum.	• Check substrate for proper treatment level. • Check ink for proper surface tension. • Check with supplier of film or foil. • Ask ink manufacturer for help. • Add wetting agent to ink.
	Ink dries too slowly.	• Add faster solvent.
	Ink viscosity is too low.	• Add more solvent. • Monitor viscosity more carefully.
	Ink viscosity is too high.	• Add more solvent. • Monitor viscosity more carefully.
	Drying system is inadequate.	• Increase drying capacity. • Increase airflow on substrate. • Reduce the press speed.
Ink setoff (second impression setoff).	Ink sets off from print to roller to next print.	• Increase drying between stations. • Use faster ink. • Increase lead or festoon distance between stations. • Reduce press speed.
Ink starvation: print fades.	Ink dries on roller.	• Add slower solvent.

Troubleshooting Flexographic Ink Problems

Problem	Cause	Solution
Mottle: random patterns appear in the print.	Ink viscosity is too low.	• Add fresh ink. • Add varnish. • Maintain proper viscosity.
	Ink is contaminated.	• Strain or replace ink.
	Impression cylinder is dirty.	• Clean the press.
	Anilox roll feeds too little ink.	• Check the anilox roll. • Increase press speed. • Check impression setting.
	Plate surfaces are uneven.	• Clean or replace the plate. • Check the plate cushion or mounting material.
One-color moiré.	Screen pattern of roller is too coarse.	• Use finer screen pattern.
Patterns in print and loss of control due to foaming of ink.	Ink formulation promotes foaming.	• Change ink. • Add defoamer.
	Ink free-falls from hose to sump.	• Lower the return hose below the ink surface.
Picking: printing plates pick ink from dried print.	First-down ink is not dry.	• Add faster solvent to first-down ink. • Increase dryer temperature. • Decrease press speed. • Add slower solvent to second-down ink.
	First-down ink is skinning.	• Add slower solvent to first-down ink.
Pinholes: random voids appear in printed film.	Ink fails to wet the film (mostly on moisture-proof cellophane).	• Ask ink manufacturer for help. • Add wetting agent. • Consult the film supplier.
Pinholes: regular pattern of voids appear in printed film.	Fast-drying ink reproduces pattern of anilox.	• Add slower solvent.

Troubleshooting Flexographic Ink Problems

Problem	Cause	Solution
Poor trapping: second-down ink is not held by first-down ink.	Drying rates are improper.	• Make first-down ink dry faster. • Reduce viscosity of second-down ink.
Precipitation: ink pattern is caused by resin coming out of ink. (Ink "sours" or "kicks out.")	Solvents are unbalanced.	• Ask ink manufacturer for recommended solvent. • Add small amount of stronger solvent ("sweetener" solvent).
	Loss of ammonia or amine causes low pH.	• Add recommended ammonia or amine.
	Alcohol in the ink absorbs water.	• Add anhydrous alcohol. • Keep ink fountains covered. • Ask ink manufacturer for solvent system that is less sensitive to water.
Rub resistance or scuff resistance is poor.	Ink is improperly formulated.	• Ask ink manufacturer to reformulate ink. • Add wax compound.
Skinover, bubbles appear in dried ink film.	Surface of ink dries too fast.	• Add slower solvent to ink (avoid excess fast solvent). • Add silicone or other wetting agent.
Viscosity out of control.	Souring: pH of ink is too low.	• Consult ink manufacturer.
	Foam is entrained in ink.	• Solve foaming problem.
	Solvent evaporates.	• Replace evaporated solvent. • Install viscosity controllers.

15 Letterpress Inks

Letterpress printing is rapidly being replaced by offset lithography, flexo, and gravure. The greatest problem with letterpress printing is the excessive time required to make ready a press with letterpress plates compared to the low cost and convenience of preparing offset plates. However, many old letterpress machines are still used, and new letterpress equipment is still being purchased, but only by small job shops which print labels, tags, envelopes, etc.

Letterpress and Lithographic Inks

Since the large, complex inking system on a letterpress resembles that of a lithographic press, there are many similarities between letterpress and lithographic inks. For example, both inks are long, thixotropic, paste inks.

Sheetfed letterpress inks, like most other inks for sheetfed printing, dry by oxidative-polymerization. Web letterpress inks dry by penetration or evaporation. Because letterpress printing generally applies a thicker ink film than offset printing applies, letterpress inks usually have less pigment (or color strength) than lithographic inks. The lower color strength makes letterpress inks less costly than lithographic inks.

Letterpress inks do not scum, tone, or emulsify. Ink drying problems, aggravated in lithography by water and dampening solution (which letterpress printing does not require), are greatly reduced.

Typical formulations of letterpress ink

News Ink		Sheetfed Ink	
Parts		Parts	
14	Furnace black	25	Furnace black
2	Methyl violet toner	10	Iron blue toner
74	Nine-poise mineral oil	55	Long-oil linseed alkyd
10	Half-poise mineral oil	5	Two-way paste drier
		5	Wax compound
100		100	

News Inks

The absorbency of the newsprint determines how and whether the news ink will dry. The oil, in which the pigment is dispersed, is absorbed into the sheet, leaving the black pigment more or less trapped by the fibers on the surface of the paper. Since letterpress news inks contain little or no binder, they have little, if any, rub resistance and tend to rub off the newsprint and onto a person's

Appropriate quality control tests for letterpress inks

Wet Ink Tests

Color
Drying
Density/Specific Gravity
Masstone
Fineness of Grind
Undertone
Tack
Tinctorial Strength
Length/Fly/Misting
Opacity/Hiding Power
Viscosity

Dried Ink Film Tests

Color
Drying
Scuff Resistance
Rub Resistance
Gloss
Odor

hands and clothing. Adding a binder to the ink increases its cost, an expense that newspaper publishers often doubt that their advertisers and subscribers will support. Note: inserts are often printed with low-rub heatset inks.

A typical news ink consists of carbon black and a high-viscosity mineral oil blended with a low-viscosity mineral oil to give the ink the desired viscosity. The cost of these inks per pound is much lower than the cost of other printing inks per pound. The tremendous volume makes the efficient manufacture of news inks an important business. The viscosity of news inks must be low enough to allow them to be pumped from delivery trucks to storage tanks to the pressroom and still be suitable for printing.

Colored news inks are usually referred to as "run of press" (ROP) inks. The ROP inks are the colors that have been accepted as standard process and spot colors for newspaper printing. (Contact the American Newspaper Publishers Association [ANPA] for additional information.) Color sections can also be added to newspapers by inserting colored printing produced by a commercial offset lithographic press.

Sheetfed Inks

Sheetfed letterpress inks are similar to sheetfed offset inks. However, they are not as highly pigmented because letterpress applies a thicker ink film than the one offset applies.

An offset ink that has been reduced with varnish or transparent white ink is satisfactory for use with letterpress machinery. Letterpress inks often contain an extender in the form of clay or another transparent white pigment to provide suitable body to the ink. Requirements for letterpress inks are usually less stringent that those for offset inks.

Troubleshooting Letterpress Ink Problems

Problem	Cause	Solution
Gloss mottle: gloss varies and mottle appears.	Paper absorbency varies.	• Increase ink viscosity to increase holdout. • Decrease ink viscosity to decrease holdout. • Replace the stock. • Add talc to dull the ink film.
Mechanical ghosting: a ghost appears on the same side of the press sheet.	Ink is too thin.	• Add varnish to ink.
	Printing plate does not have enough ink.	• Increase ink feed.
	Ink is too transparent.	• Have ink manufacturer make ink more opaque.
	Inking system is out of adjustment.	• Measure and correct rollers and adjustment.
Mottle: unwanted random variations appear in gloss or color.	Variations in paper absorbency affect holdout of ink or varnish.	• Increase ink body to increase holdout. • Decrease ink body to decrease holdout. • Replace paper or board. • Increase ink feed.
Picking: ink pulls bits of paper from the paper surface.	Ink is too tacky.	• Reduce ink tack. • Have ink manufacturer furnish tack reducer. • Change stock.
Scuffing: print becomes scuffed.	Dried ink film is too brittle.	• Add wax to ink. • Have ink manufacturer reformulate ink.
	Paper is abrasive.	• Select a softer paper.
Setoff: printed ink film transfers to back of adjacent paper.	Ink sets too slowly.	• Replace with quicker setting ink. • Increase spray powder. • Reduce ink feed. • Stack small lifts.

Troubleshooting Letterpress Ink Problems

Problem	Cause	Solution
Setoff: printed ink film transfers to back of adjacent paper *(continued)*.	Ink is too soft.	• Choose proper ink. • Add body varnish. • Have ink reformulated.
	Paper holds out ink excessively.	• Formulate ink to match the sheet. • Change stock.
Sheetfed ink dries too slowly.	Drier is insufficient.	• Add drier.
Show-through: image shows through the sheet.	Paper is too transparent.	• Change to a more opaque stock.
Strike-through: ink penetrates the sheet.	Ink is too soft.	• Replace the ink. • Reduce ink feed. • Add body varnish. • Change to a heavier stock.
	Pressure is excessive.	• Reduce packing.
Variations in color result from variations in ink or ink feed.	Ink body varies.	• Be sure all additives are thoroughly mixed before placing on press. • Roll out sample of ink on stock before going to press.
	Ink color varies.	• Be sure ink is thoroughly mixed before going to press. • Confer with ink manufacturer.
Variations in dry trap: dried ink film does not properly retain wet ink from press.	Ink crystallizes.	• Decrease spray powder. • Ask ink manufacturer for trapping compound. • Ask ink manufacturer to supply inks that will not crystallize.

Troubleshooting Letterpress Ink Problems

Problem	Cause	Solution
Variations in dry trap: dried ink film does not properly retain wet ink from press *(continued)*.	Excessive wax is in dried ink film.	• Avoid hard wax in first-down ink. • Work with ink manufacturer to be sure inks meet needs.
Variations in wet trap: second-down ink is not properly retained on printed sheet.	Ink film thickness varies.	• Measure and control ink film thickness on press.
	Ink tack varies.	• Work with ink manufacturer to control tack.
Web ink dries too slowly.	Solvent remains in ink.	• Have ink manufacturer reformulate ink. • Increase dryer temperature. • Have ink manufacturer use faster solvent.

16 Screen Inks

Screen process printing, which dates back to Egyptian and Chinese textile making in 700 A.D., was initially called "silk screen" because the first screens were made from silk. Silk is still used sometimes, but nylon, polyester, steel, or other metals are used more often. The term "serigraphy," which means silk writing, had been proposed to replace the term screen printing, but it never became popular.

Even though screen printing is almost always carried out on single objects (sheets, bottles, etc.), web screen presses are also used. Web screen presses are needed primarily for long runs and obviously require very fast-drying inks.

One of the most important characteristics of screen printing is its diversity. This makes it especially suitable for short runs (1,000, 100, or even fewer). The wide variety of applications requires a correspondingly wide variety of ink formulations. The extensive variety of substrates printed, products produced, and materials applied by screen is illustrated by the examples listed in the accompanying tables.

Examples of substrates printed by screen	Paper Paperboard Steel Aluminum Glass Ceramics Wood Leather Decorative laminates Masonite	Plastics Polyethylene Polypropylene Polystyrene Poly(vinyl chloride) Nitrocellulose Polyester (Mylar)	Textiles Cotton Canvas Denim Wool Acrylic Nylon Polyester
Examples of products produced by screen	Posters and displays Greeting cards T-shirts Highway signs Nameplates Clock faces	Printed circuits Plastic bottles Glass bottles Food packages Perfume bottles Fine art serigraphs	Awnings Calendars Wallpaper Shower curtains Floor covering Apparel
Examples of materials applied by screen	Japanese lacquers Floor adhesives Vitreous enamels	Heat transfer inks Scratch-off inks Decorative coatings	Protective coatings Fluorescent inks Etch resists

The Nature of Screen Inks

Screen inks are often referred to as "paints" (especially by sign makers and display printers) because of their similarity to some sign paints and because they are often applied at thicknesses far greater than most printing inks.

However, their formulations indicate that they are correctly called "inks."

Because of the diversity of products, the volume for any one screen ink is limited. Even more than sheetfed litho inks, screen inks are specially formulated for a particular job. Inks may even be formulated differently to print the same job on different presses: a fast press requires a somewhat different ink than a slower press.

Screen printing applies the thickest film of any common printing process, making it excellent for fluorescent and light-resistant inks. The thick film applied by a screen is often one of its advantages, but as a result of technology, screen printing can also apply a thin film—one as thin as or thinner than the film applied by rotogravure.

Screen inks differ from other printing inks in another important way. To transfer well from roll to roll, most printing inks must be "long," i.e., they must show some tendency to form a string when pulled away from a wet surface. Screen inks do not have to transfer from one roll to another. Therefore, they are short and "buttery." Short inks pass through the openings of the screen without leaving fuzzy edges. If the inks were long, they would form strings when the screen was lifted from the wet print; these strings would ruin the print.

Most screen inks contain volatile solvents, which represent up to 70% of the formulation. The most common choice is a mixture of ethylene glycol monoethyl ether, commonly referred to as Cellosolve® a trademark of the Union Carbide Corporation, and mineral spirits (aliphatic hydrocarbon) blended to give the right volatility, or evaporation rate. These mixtures are called "cosolvent." Other solvents include cyclohexanone, ethylene glycol monoethyl ether acetate, isophorone, ethylene glycol monobutyl ether, high-boiling aromatic solvents, and water.

Like other sheetfed inks, oil-based screen inks that dry by oxidative polymerization are normally alkyds based on linseed or another drying oil. Ethyl cellulose, acrylics, vinyls, vinyl dispersions or plastisols, epoxy, nitrocellulose, and urethane resins are also used in making screen inks. Catalytic-curing inks are often employed for printing bottles or circuit boards.

Most screen inks dry by evaporation: high-velocity, hot-air dryers, wicket dryers, simple drying racks, flame dryers, and even microwave dryers are used.

Ultraviolet (UV) drying systems solve one of the screen printer's greatest problems, slow ink drying. Continuing development of UV-curing technology has led to the manufacture of a broad range of UV-curable inks for a variety of applications. UV inks are used regularly to print plastic bottles, containers, point-of-purchase displays, pressure-sensitive decals, printed circuits for the electronics industry, and membrane switches for the automotive industry.

Representative screen ink formulations are shown in the following table.

Typical formulations of screen inks

For Polyethylene Bottles	**Fluorescent Poster Ink**
Parts	Parts
6 Toluidine red	46 Fluorescent pigment
10 Titanium dioxide	4 Ethyl hydroxyethyl cellulose
82 Long-oil epoxy	17 Rosin ester
2 Two-way drier	29 VM&P naphtha
— Antifoam	4 Butyl Cellosolve
100	100

Selecting the Ink

Resin-based inks provide many advantages and are now widely used. They are relatively inexpensive, provide good coverage and high color strength, and give short inks when they are highly pigmented. Extenders such as clay or calcium carbonate, commonly included to provide proper rheology without giving excessive color strength, also reduce the cost of the ink. Because the ink film is so thick, high color strength is ordinarily unnecessary. Extenders also reduce the tendency of the ink to bubble and form pinholes during drying.

For printing on glass bottles, the "ink" may be a dyed or pigmented plastic that cures on drying, or it may consist of a pigment (an oxide colorant) added to a glass frit that fuses with the glass on annealing. These glass frits or vitreous enamels can also be applied to aluminum or other substrates.

Transparent pigments, used alone, tend to highlight any imperfections in the mesh or squeegee by accentuating any variation in thickness. If transparent pigments like phthalocyanine blue or diarylide yellow are used, an opaque pigment such as titanium dioxide is also added.

Altering the Ink

Because of the small lots and wide varieties of ink formulations used in screen printing, the printer must often alter the ink to make it perform properly. Before adding anything to dilute or thin an ink, the printer should consult the ink manufacturer. In fact, the best time to ask the supplier about additives is when ordering the ink. That way, appropriate solvents and varnishes can be ordered at the same time.

A retarder slows drying. It is a high-boiling (low-volatility) solvent that is added to an ink to make it dry or evaporate more slowly. To speed drying, or reduce an ink's drying time, one adds a volatile (or "fast") solvent. If oxidative-drying inks are used, a drier is added to speed drying. The ink must dry quickly, but not so fast that it dries in the screen.

A solvent will normally reduce the viscosity of an ink, making it flow more readily. To increase the viscosity, or body, of an ink, a suitable varnish must be added. The viscosity of the ink must be low enough to permit the ink to flow through the screen, but not so low as to smear, slur, or give a muddy print.

Ink Mileage

Ink mileage is affected by the substrate and by the size of the mesh. Rough, absorbent surfaces require more ink than smooth, impervious surfaces; coarse screens deposit a thicker ink film and give lower mileage than fine screens.

Selecting the Screen

After the ink has been chosen, the printer must select a proper screen. The material chosen for the stencil and for blocking must resist the ink, and the mesh must deliver an ink film of the proper thickness. Selection of the screen fabric must take into account the nature of the ink.

The fabric may be blocked out in the nonprinting area with lacquer, shellac, or glue (liquid blockout method), with a paper or film (hand-cut blockout method), with tusche and glue (tusche-glue resist), or by photo stencil methods.

Fabrics are identified by material (nylon, stainless steel, polyester) and by diameter of the filament. Multifilament fabrics are designated by a number followed by xx, which specifies the strength of the weave, for example 12xx or 15xx. Monofilament fabrics are identified by the number of threads per centimeter or inch, example: 180, 230. In addition, the letter S signifies "small shaft" (narrow diameter), T signifies "intermediate," and HD signifies "large shaft." A 230HD fabric will apply more ink than a 230S because of the direct relationship between ink film thickness and thread thickness.

Ink film thickness, opacity, and drying rate are all affected by the fabric chosen. Fluorescent and metallic inks require a coarse-mesh screen to prevent clogging with the coarse pigments.

The relationships between multifilament screen designations and mesh counts are given in the accompanying table.

Mesh counts

Multifilament Designation	Meshes per Linear Inch	Multifilament Designation	Meshes per Linear Inch
6xx	73	15xx	144
7xx	81	16xx	152
8xx	85	17xx	160
9xx	95	18xx	170
10xx	106	19xx	175
11xx	114	20xx	180
12xx	122	21xx	185
13xx	127	22xx	200
14xx	136		

Quality Control

Because of the exceptionally broad variety of products produced by the screen process, a complete discussion of quality control is impractical. As with all inks, color match, color strength, and fineness of grind are important. Adhesion to the substrate, and compatibility with screen, squeegee, and stencil material should be checked. Suitability for the proposed end use is always important. This is often determined with tests for light resistance, product resistance, weathering resistance, laundering, and the like. Some quality control tests appropriate for screen inks are listed in the following table.

Appropriate quality control tests for screen inks

Wet ink film tests	Dried ink film tests
Color	Color
Viscosity	Opacity/Hiding Power
Masstone	Rub Resistance
Length	Scuff Resistance
Fineness of Grind	Gloss
Density/Specific Gravity	Adhesion
Tinctorial Strength	Flash Point
Tack	Drying Rate
	Flexibility
	Lightfastness

**Products
Printed by
Screen**

Decalcomanias. Pressure-sensitive decals or waterslide decal transfers are usually printed by screen because the process delivers thick, opaque films with enough flexibility to withstand the movement of the carrier paper while they are being transferred. These inks usually require good light resistance. UV inks have been used to successfully screen-print pressure-sensitive decals.

Circuit boards. When a thick film is required on a printed circuit, screen process is often the best way to print it. The ink must adhere to clean copper and resist the chemicals used in etching the copper to produce the circuit. If it is necessary that the ink be removed with a solvent or alkali after etching, the ink must be sensitive to solvent or alkali.

Posters. Posters are printed with poster inks on a variety of board and paper stocks; they may be clay-coated, patent-coated, or liner. Clay-coated board is well suited for photographic halftone printing. Nonoxidizing resins and oxidative drying inks are used the most. Overprinting with a gloss varnish extends the life of the print.

Metals. Enamel inks formulated from oil-based alkyds modified with melamine or urea formaldehyde, cellulose lacquer, epoxies, and other synthetic resins yield attractive signs for outdoor use.

The metal surface must be thoroughly degreased; aluminum is often anodized or given a nitrocellulose wash for the ink to adhere well. Baking enamels yields a product that is tough and has good resistance to aging, light, and weather.

Even more permanent are vitreous-enameled aluminum and steel products. Vitreous enamels are glasslike material or frit ground together with oxide colorants, clay, and water. After degreasing and surface treatment of the aluminum, it may be screen-printed with enamel based on borosilicates and immediately fired (without drying) at very high temperatures.

Plastics. Pigments for plastic printing must not migrate or bleed into the plastic. The solvent must be able to etch the plastic enough to improve adhesion without causing crazing (stress cracking of the plastic surface).

Thermoplastic adhesives or binders are helpful if the plastic is to be vacuum-formed after printing.

The awkward shapes of polyethylene bottles are readily screen-printed, and the thick ink deposits provide glossy and bright colors.

Glass. Inks for glass are either enamels or frits that are fired at high temperatures, or epoxy or other plastics that are baked at lower temperatures. Oil-based and synthetic resin-solvent-based inks are used to print items like dials, mirrors, and glass signs. UV inks are used to decorate mirrors.

Textiles. Plastisols and emulsions are the two kinds of inks commonly used to print textiles. Inks based on an acrylic emulsion are suitable for all types of fabric and are printed directly onto it. They will dry at room temperature, but to achieve resistance to laundering, they must be cured 2 or 3 min. at 320°F (160°C). A plastisol is a (dry) vinyl resin dispersed in a plasticizer; there is no solvent. The plastisol is pigmented and printed on the fabric. When heated above 300°F (149°C), the plasticizer "fuses with" the resin and a film is formed. Since the plastisol penetrates the fabric, the film formed on heating incorporates the fabric, producing an excellent bond. Plastisols can also be printed onto release paper, partially cured, transferred to the fabric, and then cured completely.

Trouble-shooting

In order for the ink manufacturer to help the printer solve ink problems, the printer should be ready to supply the following information:
- Screen material (e.g., monofilament, nylon)
- Mesh size (e.g., 160)
- Stencil material (e.g., blockout materials, diazo-sensitized photoscreens)
- Squeegee composition (e.g., polyurethane)
- Squeegee hardness and durometer (e.g., hard—65–75 Shore A)
- Needed coverage (e.g., 800–1,200 sq.ft./gal., 25–38 m²/L)

Troubleshooting Screen Ink Problems

Problem	Cause	Solution
Bubbles appear in film.	Print stroke foams ink.	• Add thinner. • Add silicone or other antifoam compound.
Ink dries in screen.	Ink dries too fast.	• Add retarder or "slow" solvent.
Ink dries too slowly.	Ink contains insufficient dryer.	• Add more drier.
	Solvent is too slow.	• Add faster solvent.
	Ink is too thick.	• Change screen fabric. • Reduce ink viscosity.
Ink film is too transparent.	Pigment is too transparent.	• Use another pigment. • Add TiO_2. • Print thicker film.
Ink film is too thick.	Ink is too viscous.	• Add retarder or thinner.
Ink is stringy.	Ink is too long.	• Have ink reformulated. • Add clay or transparent white pigment to ink.
Poor image quality: image is blurred or smears.	Ink viscosity is too low.	• Use a stiffer ink. • Add a stiffer varnish. • Print thinner ink film with a finer-mesh fabric.
Substrate puckers.	Ink solvent attacks film.	• Reformulate ink with a poorer solvent. • Apply a thinner ink film.
Waves or ridges appear in print.	Air velocity is too high.	• Reduce air velocity. • Print thinner ink film.

17 Toners and Specialty Inks

Toners

Ink manufacturers define toner as the pigment used to mask or alter an undesired color in printing inks, such as the blue toners used to mask the brown tones in black inks. However, in this chapter, toner is defined as the colorant used in electrostatic printing. Toners are not usually considered to be printing inks, but, like printing inks, the toners used in electrostatic printing are formulated with pigments, dyes, and resins. Toners can also be any color, but most of them are black. Used with electrostatic duplicators, they produce an image that is fixed by heating. This means that they must have the right electrical properties to stick to the duplicator drum and transfer to the paper and they must be thermoplastic, i.e., they must melt or soften when heated.

Toners are either dry or wet. Wet toners—dispersions of the toner in a hydrocarbon oil—are used with zinc-oxide-coated copier paper or with a process called **liquid-toner transfer.** The wet toners used on zinc oxide paper have largely been replaced with dry toners that are used on plain-paper copiers. Dry toners are also being used in place of liquid-toner transfer.

In dry electrostatic printing, a semiconductor drum or plate is charged in the dark. Selenium metal is commonly used as a semiconductor, but organic semiconductor polymers are used in some copiers. The charge disappears (is conducted away) in the presence of light, and when the toner is applied, it sticks to the unexposed, charged surface. The surface can be exposed with light reflected from a print, or it can be exposed with a laser beam controlled by a computer. In any case, the image is created very rapidly, allowing the machine to print several impressions per second. (Some of these machines print 10,000 impressions/hr.)

The technology for toners is developing very rapidly. Electrostatic printers and laser printers are used for desktop publishing, which includes the production of training manuals, maintenance and repair manuals, short-run books, and other short-run commercial printing.

Most dry electrostatic toners are prepared from polymers that soften but do not really melt when they are heated so that they retain a sharp impression on the print. Intensive mixing equipment is required to combine them with pigment during manufacture. During mixing, the toner melts to form a tough solid that must be chilled, pulverized, and then pulverized again. Other materials, such as flow promoters, charge modifiers, and drum cleansing agents, may be added to the dried powder. Dry toner forms a permanent image. It is solvent-free and therefore nonpolluting.

Duplicator Inks

A variety of inks are used on offset duplicator presses. Since these are offset inks, much of Chapter 12, "Lithographic Inks," is applicable. Rubber-based inks are very popular for duplicators. They dry by absorption and coalescence when they are printed on bond or other uncoated paper, but remain open on the press for days at a time (see "Cyclized Rubber" in Chapter 6). When printed, the oil from rubber-based inks drains into the sheet, and the rubber forms a film that holds the pigment and adheres reasonably well to the sheet.

Rubber-based inks used to be notorious for misting. The rubber formed tiny bands at the exit nip of the press. When these bands broke, droplets of ink flew into the air.

Paper dust, debris, and any other dust in the air accumulates in the ink and destroys its ability to print cleanly and sharply if the press is not cleaned regularly. For good printing, the press should be cleaned at least once a day.

Ink Jet Inks

Ink jet printing inks must be free from particles and must not dry in the jet. The tiny nozzles in ink jet printers are easily clogged. To avoid clogging, the ink is made with a dye instead of a pigment. The solvent must be "slow" or have low volatility to avoid evaporation and clogging of the jet. Ink jet inks resemble fountain pen or writing inks. They consist of dyes in water or glycol.

Magnetic Inks

Magnetic inks are used for printing account numbers and other information on bank checks so that they can be sorted and handled by magnetic ink character recognition (MICR) devices. Magnetic iron oxide, which is the pigment in magnetic tapes, is also the pigment in magnetic inks. Magnetic iron oxide does not make a good ink, and the ink manufacturer must be skilled enough to produce an ink that will flow, print, dry, and be easily read by MICR devices. The printer must not alter this ink in any way because additives affect the magnetic strength of the image and the way it reads.

Printed Circuit Inks

Printed circuit boards usually consist of a layer of copper on an insulating material. An ink is applied by screen printing in the design of the circuit, and when dry, the ink acts as a resist, i.e., it protects the printed area from etching. The printed copper is then etched and removed, except where it is protected under the hardened resist (ink). Very fine detail and complicated circuitry can be obtained using this method. It is also possible to form a circuit board by applying conductive silver inks by screen printing.

Optical Character Recognition Inks

Any ink can be read by an optical character recognition (OCR) device provided the reading device is sufficiently sensitive. Bar-code readers are the most familiar OCR technology. OCR is useful not only in sorting documents—checks, invoices, and mail—but also in converting printed or typed matter into digital signals for the computer and for typesetting.

For other systems, such as the Universal Product Code (UPC), any sharp and clear, optically dense ink can be detected by the reader.

Scratch-off Inks

To hide the numbers on lottery tickets or instant winner cards, an opaque ink that adheres well yet comes off easily when scratched with a coin, fingernail, or other object is required.

The inks are usually based on rubber that contains vehicles and a highly opaque pigment, such as aluminum powder.

Water-Dissolvable Inks

Some lottery tickets are four- or five-part structures that conceal the symbols and obliterate any pattern that would disclose them. They can be printed using a protein resin usually dissolved in glycol and colored with an opaque pigment. The inks dry rapidly but can easily be redissolved with a wet towel to disclose the number or word printed underneath.

To ensure that the number cannot be detected without washing away the top ink, it is printed as a light screen and two layers of water-washable ink (for example, aluminum followed by black) are printed over it.

Because the ink is water-washable, it cannot be printed by lithography. It can be printed by screen, letterpress, or offset letterpress ("dry offset").

"Scratch-and-Sniff" Inks

"Scratch-and-sniff" inks contain perfume or scent encapsulated in minute gelatin sacs that are broken when scratched by a fingernail. Controlling printing pressure is critically important when printing these inks or the capsules will break.

Autoclave Inks

Autoclave inks contain dyes or pigments that change color when subjected to high temperature, pressure, and/or steam. They are used chiefly to print packages containing medical items that require sterilization. The color of the printed ink immediately shows whether the package has been autoclaved. These inks can be printed by any process. Thermal indicating inks that show temperature changes are also available. Some are reversible, some are not.

Water-Developing Inks

Children's coloring books are sometimes printed with an ink that develops color when stroked with a wet brush. The book is printed with a black ink that contains a water-soluble dye, using a fine halftone screen. The ink dries hard, but the dye is activated by water from the wet brush. Colors can be combined to give several effects. Because of the water-sensitivity, the books cannot be printed by offset lithography.

Invisible Inks

There are several varieties of invisible inks. One consists of clay-pigmented ink which, when printed on an enamel paper, forms an image that is nearly invisible when the ink has dried. The clay is more sensitive to a pencil mark than the enamel paper, so rubbing with a pencil makes the design printed with the clay apparent.

Transparent inks containing materials that fluoresce under "black light" (ultraviolet) are invisible in ordinary light. This ink is used on horse and dog race tickets, prize coupons, and even stock certificates to guard against fraud. The clerk who gives winners their money can easily check the validity of the ticket by placing it under a black light to read the "invisible" security design.

Other invisible inks can be made visible by treating them with chemicals or vapors that react with an ingredient in the ink to develop a color.

Security Inks

Security inks are printed in the background on checks and other negotiable documents. The amount of money is written over the printed security check. Any attempt to erase or alter the figures will destroy or smear the ink. These inks are usually water-soluble, which means that they will dissolve easily upon contact with a liquid ink eraser or chemicals in ink eradicators.

Glossary

abrasion test A test designed to determine how much rubbing and scuffing a dried ink film can withstand.

agglomerate A cluster of undispersed ink particles.

alkali blue Pigment dispersed in lithographic varnish and used as a toner in carbon black inks. *Alternative term:* reflex blue.

antioxidants Agents that retard the action of oxygen in drying oils and other substances subject to oxidation.

antiskinning agent An antioxidant that prevents sheetfed offset inks from skinning over in the can.

ASTM The American Society for Testing Materials.

Bingham plastic A liquid that will not flow at all until a finite force is applied.

black The absence of color; an ink that absorbs all wavelengths of light.

blown oil Product obtained by blowing air through heated drying or semidrying oils.

body The viscosity, consistency, length, or flow of an ink or varnish.

brightness Percent of reflectance at a standard wavelength (457 nm in the standard test—a blue wavelength that most readily detects the yellowing of paper).

cadmium yellow An inorganic pigment that is principally cadmium sulfide and cadmium selenide.

carbon black A black pigment that consists mostly of elemental carbon, a small percentage of ash (mineral matter), and a somewhat higher percentage of volatile matter.

catalyst An ingredient added to an ink formulation (usually a small amount is sufficient) to bring about a chemical reaction between other ingredients. (A drier is a catalyst.)

Cellosolve®   A trademark of the Union Carbide Corporation for ethylene glycol monoethyl ether, a relatively slow-drying flexo ink solvent frequently used as a retarder, usually in amounts less than 10%.

china clay   Hydrated aluminum silica, which is used in gravure and screen printing inks and as an extender in letterpress inks. *Alternative term:* kaolin.

chrome green   A mixture of chrome yellow and iron blue.

chrome yellow   A yellow pigment that is principally pure lead chromate.

cobalt drier   A material containing chemically combined cobalt used to accelerate oxidation and polymerization of an ink film.

color strength   The intensity of a color—i.e., the concentration or dilution (attenuation) of a color—or the relative amount of pigmentation in an ink film. *Alternative terms:* chroma; intensity; saturation; and tinctorial strength.

cyan   A pigment that absorbs all of the red wavelengths of light but none of the blue or green; incorrectly called "process blue."

deinking   A treatment, usually with caustic, given to wastepaper to soften the ink and remove it from the paper fibers so that the paper fibers can be recycled.

diluent   A solvent that is added to reduce viscosity; a nonsolvent or poor solvent for a vehicle that also reduces the viscosity of the vehicle.

dispersibility   The degree to which pigment particles can be separated and surrounded by vehicle.

drawdown   A film of ink deposited on paper by a smooth-edged blade to evaluate the undertone and masstone of the ink.

dried pigment   A pigment in dry or powder form. *Alternative term:* dry color.

drier   A catalyst that promotes ink drying (the conversion of a wet ink film to a dry ink film).

drier dissipation	A loss in catalytic power of a drier due to a physical absorption or a chemical reaction with certain pigments.
dryer	A mechanical device designed to accelerate the drying of inks.
drying time	The time required for an ink to form a tackfree surface after being applied to the paper, or other printed surface.
electro-magnetic spectrum	A continuous sequence of wavelengths of electromagnetic energy, ranging from very long radio waves to extremely short gamma rays.
electrostatic printing	A printing method in which a drum or flat plate, usually made of selenium or a photoconductive polymer, is charged electrostatically (in the dark). When light, either from reflected copy or from a computer-controlled laser beam, hits the drum, the charge is dissipated leaving an electrostatic image. Toner is then applied to the drum or surface. It is retained by the electrostatic image, transferred to the paper, and then fused to produce a permanent copy.
extender	A white pigment used with a colored pigment either to reduce its strength or improve its working properties.
fineness of grind	The degree of dispersion of a pigment in a printing ink vehicle, usually measured on a grindometer or grind gauge.
flash point	The lowest temperature at which a substance gives off vapor that will ignite when exposed to a flame.
flexography	A method of direct rotary printing from resilient relief plates (rubber, synthetic rubber, plastic, photopolymer, and synthetic polymer), which carry fluid inks to virtually any substrate.
fluid ink	See *ink, fluid.*
fluorescent pigment	Fluorescent dyes dispersed in inert, insoluble resins and ground to a small size.

flushed pigment   The result when a wet pigment is processed in a mixer along with a selected varnish, the pigment becoming preferentially wet with the varnish and transferring from the water to the varnish.

gloss ghosting   The transfer of a printed image (but not setoff) from the front of one sheet to the back of another (but through it) caused by the chemical-activity influence that inks have on each other during their critical drying phases. *Alternative terms:* chemical ghosting; fuming ghosting.

gravure   Printing from an inked image area that is etched into the surface of the image carrier.

gum blinding   The result when an image becomes wet with water, or when gum is deposited on the image, and the ink fails to adhere to the image.

heatset inks   Letterpress and lithographic inks that dry primarily by evaporation (absorption and oxidation may also be involved) as the web goes through a hot-air dryer.

hickey   Small solid area sharply defined and surrounded by a white halo on the print.

hue   A visual property determined by the dominant light wavelengths reflected or transmitted.

hydrocarbon   Organic substance containing only carbon in combination with hydrogen.

ignition point   The temperature at which the vapor-air mixture given off by the liquid continues to burn after spontaneous combustion.

inhibitor   A compound (usually organic) that retards or stops a chemical reaction, such as corrosion oxidation or polymerization. Such substances may be regarded as negative catalysts. Antioxidants are inhibitors.

ink, fluid   An ink having a low-viscosity vehicle. *Alternative term:* liquid ink.

ink, paste   An ink having a high-viscosity vehicle.

ink jet printing — Printing method in which a jet of electrostatically charged ink droplets are projected onto the substrate as an image.

ink setting — The increase in viscosity or body (resistance to flow) that occurs immediately after the ink is printed.

kaolin — See *china clay*.

lake — The pigment that results when a soluble dye is converted into a pigment in the presence of an inorganic white base such as alumina hydrate or white gloss.

length — The length of the string or filament that an ink will form before breaking.

letterpress — A method of printing from an inked image area raised above the nonimage area on hard plates using paste inks.

lightfastness — The extent to which a material—e.g., ink or paper—is resistant to the action of light.

liquid drier — Metal salts suspended in a liquid, such as a petroleum solvent.

lithography — Printing from a planographic, or flat, surface in which image and nonimage areas are chemically different.

magenta — A color that absorbs all of the green wavelengths of light, but none of the red or blue; incorrectly called "process red."

magnetic black — Black iron oxides used as the pigment in black inks for magnetic ink character recognition.

manganese driers — A material containing chemically combined manganese used to accelerate the oxidation and polymerization of an ink film.

mechanical ghosting — A ghost image that is always carried on the same side of the sheet and caused by inadequacies in lithographic and letterpress inking systems, e.g., ink starvation, or a depressed area of the blanket caused by use on another job.

metamerism   The phenomenon of colors matching under one light but not under a different light.

molybdate orange   Inorganic pigment resulting from the coprecipitation of lead chromate with lead sulfate and lead molybdate.

mottle   Irregular and unwanted variation in color or gloss caused by uneven absorbency of the substrate.

offset lithography   A planographic printing process in which the image on the printing plate is transferred to an intermediate surface, a blanket, before being printed on the substrate. See *lithography*.

pH   The potential of the hydrogen ion, a measure of the degree of acidity or alkalinity, expressed as the negative logarithm of the concentration of hydrogen ions in moles per liter.

phloxine   A brilliant bluish red (magenta) pigment widely used in letterpress and news inks but not lithographic inks because of its tendency to bleed in alcohol and water.

phthalo-cyanine   Blue and green ink pigments characterized by extreme lightfastness and resistance to solvents, acid, and alkali. The blue is now widely used in process inks.

picking   A disturbance of the paper's surface that occurs during ink transfer when the forces required to split an ink film are greater than those required to break away portions of the paper surface.

pigment   Finely ground solid material that gives color to an ink.

piling   Accumulation of pigment or coating from the paper onto the blanket, plate, or rollers.

piling ghosting   A ghost of an image that appears on the reverse side of coated paper printed on a blanket-to-blanket press (usually a web offset press) caused by the uneven pressure that has resulted from piling on the blanket adjacent to the image.

plasticizer   An ink additive that makes ink softer, more flexible, and more adherent to the substrate.

polymeri-zation
A chemical reaction (usually carried out with a catalyst, heat, or light) in which two or more relatively simple molecules (monomers) combine to form a chainlike macromolecule or polymer.

red lake C
A warm, bright red pigment used in printing inks. It is chemically similar to the rubine pigments.

reducer
An ink additive that softens and reduces the tack of the ink.

relative humidity
The amount of moisture present in the air, expressed as a percentage of the amount of moisture required to saturate the air at a given temperature.

resinated pigment
A pigment that has a surface treated with a suitable resin to make it more easily dispersible.

rheology
The study of the flow of fluids (liquids or gases).

rhodamine
A class of clean, blue shade organic red pigments that are more magenta than the rubines.

ROP color
Process or spot color printed during the run of a newspaper; abbreviation for "run of press" color or "run of paper" color.

rubine
An organic pigment with a shade somewhat redder than a "true" magenta.

screen printing
Printing method in which ink is forced through openings in a screen and onto a substrate.

specific gravity
The ratio of the weight of a given volume of a material, e.g., solvent, to the weight of the same volume of water at a given temperature.

Stefan's equation
An equation that shows that the force required to split a thin ink film on a press is related not only to the ink body but also to the speed of the press and the area and thickness of the ink.

substrate   The surface that receives the printed image, including materials such as paper, paperboard, glass, plastics, and metal.

tack   The force required to split a thin ink film in terms of a number obtained from the Inkometer or other tack-measuring device; the sticky or adhesive quality of an ink.

thinner   A solvent for the vehicle.

thixotropy   The characteristic of a paste ink or plastic material to lose viscosity or body when being stirred or otherwise worked.

through drier   A less-active drier that dries the ink film throughout without forming a hard surface.

titanium dioxide   A brilliant white, opaque pigment used extensively to provide an opaque background for metal decorating and for printing on flexible packaging.

toluidine red   A relatively low-cost red pigment with good lightfastness but poor bleed and heat resistance.

toner   The pigment that results when a soluble dye is converted to a pigment in the absence of an inorganic white base; the colorant used in electrostatic printing.

top drier   An active drier that gives a very hard surface to the ink film.

ultramarine blue   A bright transparent blue pigment produced from sulfur, silica, china clay, or carbon (rosin pitch or charcoal) and either soda ash or sulfate salts.

varnish   The major component of an ink vehicle, consisting of solvent plus resin or drying oil.

vehicle   A liquid composed of a varnish, waxes, driers, and other additives that carries the ink colorant (pigment), controls the flow of the ink or varnish on the press, and, after drying, binds the pigment to the substrate.

viscosity   Resistance to flow.

visible spectrum   The range of wavelengths of the electromagnetic spectrum—from about 400–700 nm—that affect human vision.

wax   An additive that improves slip and resistance, prevents setoff, and reduces ink tack.

wettability   The ease with which pigments can be wet by the ink vehicle.

wetting agent   An additive that promotes dispersion of pigment in ink varnish.

wet trapping   The ability of a printed ink film to accept a succeeding ink film applied on the press.

yellow   A color that absorbs all of the blue wavelengths of light but none of the red or green.

yield value   The finite amount of force needed to start an ink flowing.

Index

Abrasion resistance, testing for 131–132
Absorption 88–89
Accelerators 84
Acrylics 69
AD-LITHO color system 23
Adhesion, tests for 132–133
AdPro color system 23
Alcohols 74
Alkyd resin vehicles 62–63
Alkyds 63
ANPA-COLOR 23–24
Antisetoff compounds 78–79
Antiskinning agents 79
Aqueous inks, defoamers 80
ASTM tests,
   density and specific gravity 129
   drying 115
   gloss 114
   viscosity 124

Bar or rod viscometers 123
Bead mill 100
Bingham plastics 43
Blankets, quick release 53
Block resistance, test for 133–134
Blown oil 61

Carbon black pigments 33–34
Catalytic polymerization 93
Cellulosic resins 64–65
Chromaticity diagram 20–21
Color,
   additive and subtractive 21–22
   illumination 17–18
   ink specifications for 140
   instrumental systems 24
   matching 22–24
   measurement 18–21, 112
   mixing 24
   testing strength 113–114
Color variation 24–25
Colored inorganic pigments 38–39
Colored organic pigments 34–37
Colorimetry 20–21

Conical ink agitator 43–44
Cup viscometers 122–123

Dampening solutions, pigment bleeding in   128
Defoamers for aqueous inks 80
Deinking 94–95
Densitometry 18–20
Density and specific gravity, testing for   129–130
Diarylide yellow 34–37
Diluents and thinners 71
Disperser, twin-shaft 103
Dispersibility, pigments 29
Dot gain 25
Driers, 80–86
  accelerators 84
  chain reaction 80–82
  effect of temperature and acidity 84–86
  liquid 83
  metal 82
  mixing 84
  paste 83
Driographic inks 151
Dry trap 160
Drying,
  absorption 88–89
  ASTM tests 115
  catalytic polymerization 93
  evaporation 89–91
  heatset dryer problems 154
  infrared radiation 93–94
  oxidative polymerization 91–92
  radiation polymerization 92–93
  slow 158–159
  tests of 115–117
Drying and setting 87–95
Drying oils 59, 61–62
Drying systems, radiation 148–149
Drying time on paper, ink specifications for   141

Electron-beam (EB) vehicles 59
Electrophotography 8–9
Emulsification, 158
  ink specifications for   141
  testing for   126–128

Esters 76
Ethers, glycol 74, 76
Evaporation 89–91
Extenders 39–40

Fineness of grind, ink specifications for 141
Fineness-of-grind test 117–118
Flash point 71
Flash point test 130
Flexographic inks 197–210
Flexography 5–6
Flow 41–55
Fluorescent pigments 37
Flushed pigment 31

GATF Inkometer 119–121
Gellation 87
Ghosting 161–163
Gloss, ASTM tests 114
Gloss ghosting 161–162
Gloss testing 114
Gloss vehicles 58
Glycol ethers 74, 76
Gravure 4–5
Gravure inks 181–196
Gray component replacement 25–26

Hansa yellow 35
Heat-bodying 59, 61
Heat-seal resistance, tests for 136
Heatset dryer problems 154
Heatset ink components 98
Heatset oils 73–74
Heatset vehicles 57
Heatset web offset publication inks 147
Hickeys 160–161
High-velocity hot-air dryer 90
Hunter L,a,b chromaticity diagram 20–21
Hydrocarbon resins 64
Hydrocarbons 72–73

Illumination 17–18
Infrared radiation 93–94
Infrared vehicles 59

Inhibitors 83
Ink, 1
  additives 77–80
  body 41
  contamination 153–154
  driographic 151
  flexographic 197–210
  gravure 181–196
  in the nonimage area 154–155, 169–175
  inventory 15
  letterpress 211–216
  lithographic 145–180
  magnetic 150
  metal-decorating 149–150
  metallic 149
  nonheatset web offset publication and news 147–148
  rubber-based 68
  screen 217–224
  setting and drying 87–95
  specifications 12–14, 139–143
  storage 14–15
  testing 107–137
  vehicles 57–80
Ink drying time,
  effect of drier content 83
  effect of relative humidity (R.H.) 85
  effect of temperature 84
Ink film thickness 2, 50–51
Ink flow, 41–55, 125
  testing for emulsification 126–128
  testing for fly or mist 126
  testing length 126
Ink jet printing 9–10
Ink length 45–47
Ink manufacture 97–105
Ink mileage 11–12
Ink release, effect of stock 54
Ink specifications, 139–143
  checking 143
  color strength 140–141
  drying time on paper 141
  emulsification 141
  fineness of grind 141
  open time 141

   shelf life   141–142
   tack   142
   viscosity   142
   volatility   142
   yield value   142–143
Ink testing,
   abrasion resistance   131–132
   adhesion tests   132–133
   block resistance   133–134
   density and specific gravity   129–130
   drying   115–117
   emulsification   126–128
   fineness of grind   117–118
   flash point   130
   flow   125
   fly or mist   126
   gloss   114
   heat-seal resistance   136
   lamination   137
   length   126
   lightfastness   135
   odor and taste   136
   scumming   128–129
   skid resistance   134–135
   tack   119–121
   viscosity   122–125
   water resistance   136–137
   wet ink film   118
   working properties   114–130
Ink tests,
   color   111–114
   drawdown   110–111
   opacity   114
   optical properties   111–114
   sampling   108
   standardized   108
Ink trap   25
Ink/paper problems   156–158, 175–178
Ink/paper release   53–55
Inking system, lithographic   145–146
Inorganic colored pigments   38–39
Instrumental color systems   24
Iron blue   38

Ketones 68, 76

Lamination, testing for 137
Letterpress 6–7
Letterpress inks 211–216
Lightfastness,
  pigments 30
  tests for 135
Linting 157–158
Liquid driers 83
Liquid ink processing 99–101
Lithographic ink problems,
  catch-up 155
  emulsification 158
  gloss ghosting 161–162
  hickeys 160–161
  mechanical ghosting 162–163
  scumming 155
  tinting 155
  toning 155
  trapping 159–160
Lithographic inking system 145–146
Lithographic inks, 145–180
  contamination 153–154
  driographic 151
  for printing plastic or plastic-coated paper 151
  heatset web offset publication 147
  ink in the nonimage area 154–155
  low-rub 148
  low-solvent, low-odor 148
  magnetic 150
  metal-decorating 149–150
  metallic 149
  problems 152–153
  quality control tests 146
  typical formulations 146
Lithography 2–4
Low-rub inks 148
Low-solvent, low-odor inks 148

Magnetic black 38–39
Magnetic inks 150
Maleic and fumaric resins, rosin-modified 67
Mechanical ghosting 162–163

Metal driers   82
Metal-decorating inks   149–150
Metal-decorating vehicles   58
Metallic inks   149
Metallic pigments   39
Modified drying oils   62–63
Modified natural resins   64–68

Natural resins,   64
   modified   64–68
Newtonian flow   41–42
Nitroparaffins   76
Nondrying oils   64–70
Nonheatset web offset publication and news inks   147–148
Nonimpact processes   8–10
Non-Newtonian flow   41–42

Oil ink vehicles, primary   62
Oils,
   heatset   73–74, 75
   nondrying   64–70
Oiticica oil   62
Opacity,
   of pigments   29
   testing for   114
Open time, ink specifications for   141
Optical pyrometer   91
Organic pigments   30–31, 31–37
Overprint varnishes   58, 152
Oxidative polymerization   91–92

PANTONE MATCHING SYSTEM   23
Paper color,
   "bright"   25
   "warm"   24–25
   "white"   25
Paste driers   83
Paste ink processing   101–102
Petroleum solvents   73
Phenolic resins, rosin-modified   67
Phenolics   63
Phthalocyanine blue   35
Pigment,   27–40
   chips   32

dispersibility 29
dried 31
fluorescent 37
flush 31
inorganic 37–38
manufacture 30–32
opacity 29
particle size 27
properties 27–30
refractive index 29
specific gravity 29
texture 30
toners 32
wettability 29
Piling 46–47, 156–157
Plasticizers 77
Poly(vinyl butyral) 69
Polyamides 68
Press cake 31
Primary oil ink vehicles 62
Proof presses 109

Quality control tests for lithographic inks 146
Quickset vehicles 57–58

Radiation drying systems 148–149
Radiation polymerization 92–93
Red lake C 36
Reducers 79
Reflex blue 35–36
Refractive index of pigments 29
Resinated pigments 32
Resins,
  cellulosic 65
  hydrocarbons 64
  modified natural 64–68
  natural 64
  synthetic 68–70
Rheology 41–55, 122
Rhodamine 36
Rosin-modified resins 67
Rosins 66–67
Rotational viscometers 124–125
Rubber-based ink 68

Rubine   36

Sampling   108
Screen printing   7–8
Screen printing inks   217–226
Scumming,   155
   testing for   128–129
Setting   47
Setting and drying   87–95
Shear-dependent flow   41–42
Shear-independent flow   41–42
Sheetfed inks, lithographic   146–147
Shelf life, ink specifications for   141–142
Shortening compounds   79
Skid resistance, tests for   134–135
Solvents   70–76
Soybean and safflower oils   62
Specific gravity of pigments   29
Spectral reflectance curves   21–22
Spectrophotometry   20
Stefan's equation   48–52
Stiffening agents   79
Styrene-maleics   68
Surland Emulsification Method   127–128
Synthetic resins   68–70

Tack,   48–53
   ink specifications for   142
   Stefan's equation   48–52
   testing of   119–121
Tall oil   62
Temperature coefficient of viscosity   44–45
Terpenes   70
Test prints   108–111
Testing end-use properties   130–137
Testing inks   107–137
Thinners and diluents   71
Thixotropy   45, 87
Three-roll mills   104–105
Three-way driers   82
Through drier   82
Tinting   155
Toners   32
Toning   155

Top drier 82
Trapping 51–53, 159–160
Tung or chinawood oil 61–62
Twin-motion mixer 103
Twin-shaft disperser 103
Two-way driers 82

Ultraviolet (UV) vehicles 59
Undercolor removal 25–26
Urethanes 63

Varnishes,
  gloss 58
  overprint 58, 152
Vehicles 57–80
Vibrating-reed viscometers 125
Vinyls 68
Viscometers 123–125
Viscosity 41–45, 122–125, 142
Volatility, ink specifications for 142

Water resistance, testing for 136–137
Water-based gravure and flexo inks 58
Waxes 77–78
Wet trap 51–53, 159–160
Wettability, pigments 29
Wetting agents 78

Yield value 43, 142–143